Refrigeration

Questions and Answers books are available on the following subjects:

Pipework and Pipe Welding
Electric Arc Welding
Gas Welding and Cutting
Plumbing
Carpentry and Joinery
Brickwork and Blockwork
Painting and Decorating
Plastering
Refrigeration
Motor Cycles
Automobile Brakes and Braking
Automobile Electrical Systems
Automobile Engines
Automobile Transmission Systems
Car Body Care and Repair
Diesel Engines
Light Commerical Vehicles
Wooden Boat Construction
GRP Boat Construction
Steel Boat Construction
Lathework
Radio and Television
Colour Television
Radio Repair
Hi-Fi
Electronics
Integrated Circuits
Transistors
Electricity
Electric Motors
Electric Wiring

QUESTIONS & ANSWERS

Refrigeration

M. E. Anderson

revised by

R. H. Herbert

Newnes Technical Books

The Butterworth Group

United Kingdom	**Butterworth & Co (Publishers) Ltd** London: 88 Kingsway, WC2B 6AB
Australia	**Butterworths Pty Ltd** Sydney: 586 Pacific Highway, Chatswood, NSW 2067 Also at Melbourne, Brisbane, Adelaide and Perth
Canada	**Butterworth & Co (Canada) Ltd** Toronto: 2265 Midland Avenue, Scarborough, Ontario M1P 4S1
New Zealand	**Butterworths of New Zealand Ltd** Wellington: T & W Young Building, 77–85 Customhouse Quay, 1, CPO Box 472
South Africa	**Butterworth & Co (South Africa) (Pty) Ltd** Durban: 152–154 Gale Street
USA	**Butterworths (Publishers) Inc** Boston: 19 Cummings Park, Woburn, Mass. 01801

First published in 1948 by George Newnes Ltd
Second edition 1951
Third edition 1955
Fourth edition 1971
Reprinted 1975, 1977, 1978, 1979

Published by Newnes Technical Books,
a Butterworth imprint

ISBN 0 408 0043 0

Printed in England by
Butler & Tanner Ltd, Frome and London

CONTENTS

PREFACE

The use of refrigeration today is widespread and the purposes for which it is necessary or advantageous are so numerous that there must be a very considerable number of people concerned with it, one way or another, who would like to know a little about the many problems involved.

The types of plant in use are many and diverse, and their successful construction and utilisation involves a knowledge of several branches of engineering as well as an acquaintance with the various industries in which refrigeration plays a part. As a result it is not possible in a book of this size to deal other than briefly with the principal branches of the subject; nevertheless it is hoped that the questions and answers will be found of interest and of practical value to the student, the engineer and the user of refrigerating plant, as well as to the man who is just curious to know "what makes it tick".

In this, the fourth edition, questions are answered using both the International and the British system of units; this has been done because complete metrication is still some years ahead and it is felt that many readers will be happier using the British units to which they are accustomed—and, indeed, the Btu is still the unit of heat in this country.

R.H.H.

1

PRINCIPLES AND DEFINITIONS

What is heat?
Heat is a form of energy and is due to the motion of the molecules of which all substances are composed. The effect of adding heat to a substance is simply to increase the speed at which its molecules move, and thereby their energy.

What is cold?
Cold is a rather vague term used to denote a comparative lack of heat.

What is temperature?
Temperature is a measure of the intensity of heat in a substance and of its ability to pass its heat into anything at a lower temperature than itself.

How is temperature measured?
By taking advantage of one of two facts: firstly that the electrical resistance at the junction of two different metals (a thermocouple) varies according to the temperature, and secondly that the volume of a body varies with its temperature. In the first case a constant voltage is applied to the junction of the two metals, and the resulting current is measured with an ammeter calibrated in degrees of temperature; as the temperature at the junction varies, so will the current flowing and therefore the indicating hand on the dial of the instrument. The distance between the thermocouple and the indicator is immaterial, and this form of measurement is very suited to remote temperature reading.

Most other thermometers take advantage of the fact that the volume of a substance changes with temperature.

Liquids, usually mercury or spirit, can be enclosed in a glass tube. The variation of volume with a temperature change is shown by an alteration in the height of the liquid in the tube, which is itself marked with degrees of temperature.

The variation of volume with temperature is not the same with all materials and this fact is taken advantage of in the dial type of thermometer where the difference in expansion of the two parts of a bi-metal strip results in a torque being developed which can be used to operate the pointer on a circular dial. This is the principle used in the recording thermometer or thermograph.

What temperature scales are used?

Only two temperature scales are of importance, the Centigrade* and the Fahrenheit. On the former, the freezing point of water at atmospheric pressure is denoted by 0°C and its boiling point by 100°C; there are therefore 100 Centigrade degrees between these two temperature levels. On the latter the freezing point is 32°F and the boiling point 212°F, so that the difference between the two levels is 180°F. Hence a change of temperature of 1°C is the same as a change of 1·8°F.

To convert from °C to °F and vice versa, the following equations are used:

$$°F = (\tfrac{9}{5} \times °C) + 32$$

$$\text{and} \quad °C = (°F - 32) \times \tfrac{5}{9}$$

These rules apply whether the temperature is above or below freezing but, of course, a minus sign must be used below 0°.

Care must be taken when converting a temperature *difference* compared with an actual temperature. For instance a temperature *difference* of, say, 9°F would be the same as a *difference* of 5°C, whereas an actual temperature of 9°F would be the same as −12·8°C.

What is absolute temperature?

On the absolute temperature scale (which may be in either Centigrade or Fahrenheit degrees) 0° represents the lowest

*Now known as *Celsius*, the degree Celsius being the metric practical unit of temperature

attainable temperature at which the internal energy of all substances is zero. This temperature is −273·1°C* or −459·6°F; hence to convert Centigrade temperatures to °C absolute we add 273·1 and to convert Fahrenheit temperatures to °F absolute we add 459·6.

What units are used for measuring heat?

The British thermal unit (Btu) is used by engineers in the U.K. and the U.S.A.; the calorie is used in scientific work and generally wherever the metric system is in use. The Btu is the amount of heat which will raise the temperature of 1 lb of water by 1°F. There are two calories; the small calorie, or gramme-calorie, which is the amount of heat required to raise the temperature of 1 gramme of water by 1°C and the great calorie or kilocalorie (kcal) which is 1000 times as great, being the amount of heat required to raise the temperature of 1 kilogramme (kg) of water by 1°C. A frigorie is exactly the same as the kilocalorie and is used by European writers to denote the power of removing heat possessed by a refrigerating plant.

1 kcal = 3·968 Btu = 4·18 J. The joule (J) is the SI unit of heat, but it is not yet in general commercial use.

What is the latent heat of a substance?

The amount of heat which has to be added to unit weight of the substance to change its state from solid to liquid (latent heat of fusion) or from liquid to vapour (latent heat of vaporisation). It is used in overcoming intermolecular forces and no change in temperature results.

What is sensible heat?

Heat which results in a change in temperature; when unit weight of a substance is heated by 1° the gain in sensible heat is equal to the specific heat.

What is specific heat?

The amount of heat that has to be added to a substance to produce a given rise in temperature varies according to the nature of the substance. The specific heat of a substance is

* °C absolute are known as kelvins in SI units.

the ratio of the amount of heat which will raise the temperature of a given weight of it by 1° to the amount of heat which will produce the same rise in temperature in the same weight of water. It is independent of the temperature scale used. By definition the specific heat of water itself is 1.

What is vapour pressure?
Every liquid produces a vapour as a result of the molecules near its surface freeing themselves from the attraction of their neighbours and flying off into space. This vapour exerts a pressure on any containing vessel and the amount of pressure exerted by the vapour of any particular liquid depends solely on the temperature of the liquid surface; the higher the temperature the greater the pressure. For any liquid a graph can be drawn showing the relationship between this vapour pressure and the temperature of the liquid surface.

How can gas be liquefied?
When heat is removed from a gas its temperature is lowered until it reaches a value corresponding to the pressure (see above) after which further removal of heat liquefies the gas. Alternatively, an increase in pressure combined with a removal of heat makes it possible to liquefy the gas without reducing its temperature.

What is superheated vapour?
Vapour removed from contact with its liquid and at a temperature higher than that which corresponds to its pressure as indicated by the temperature-pressure-vapour relationship for that particular substance.

What is saturated vapour?
Vapour whose temperature and pressure are in accordance with the temperature-vapour-pressure relationship for the particular substance. Vapour in contact with its liquid is saturated.

What is the numerical relationship between heat units and units of mechanical work?

This can be expressed as follows:
1 Btu = 778 ft lbf. Thus, for example, the energy given up by a weight of 77·8 lb in falling 10 ft is sufficient to heat 1 lb of water through 1°F. Again, since a horsepower (hp) is the rate of production of energy of an engine doing 33,000 ft lbf of work per minute, 1 hp = 33,000/778 or 42·4 Btu/min and therefore 2,545 Btu/h. The kilowatt (kW), (not kilowatt hour) is another unit of power, and 1 kW = 1·34 hp or 3,410 Btu/h. Further, 1 kcal = 4·18 J, and 1 J/s = 1 W.

What is enthalpy?

This is a term that is often used in thermodynamic calculations and for most purposes it may be regarded as being the same as heat content, or the amount of heat contained in unit weight of a substance measured from some arbitrarily chosen condition taken as zero. This is not strictly correct and it is accurately defined by the equation:

$$H = I + PV/J, \text{ where}$$

H = enthalpy
I = internal (molecular) energy
P = absolute pressure
V = volume
J is the factor to convert heat units to work units.

It can be shown mathematically that if heat is added to or taken from a substance, the change in enthalpy equals the amount of heat added or removed, *provided that the pressure remains unchanged.* Again, it can be shown that the increase in enthalpy during *adiabatic* compression is equivalent to the work done in compression.

What is entropy?

Change of entropy, which is the only thing that concerns us in thermodynamics, is equal to the amount of heat added or removed during a reversible process, divided by the absolute temperature; if the temperature changes during the process, as it does in the case of sensible heat, the entropy change must be evaluated by dividing the process into so many small steps that the temperature may be considered constant for each step, and adding the results (integration).

What is adiabatic compression?

When a gas or vapour is compressed in such circumstances that there is insufficient time for any substantial exchange of heat between it and its surroundings, the compression is said to be *adiabatic*. It can be shown that during such a process there is no change in entropy and it is mainly this fact that makes the conception of entropy so useful. The compression of vapour in a refrigerating compressor is nearly adiabatic unless a water jacket is fitted.

What is a perfect gas?

One which behaves in accordance with the gas law,

$PV = MRT$, where

P = absolute pressure,
V = volume,
M = mass,
T = absolute temperature, and
R is a constant for the particular gas depending on its molecular weight.

The volume is directly proportional to the temperature and inversely proportional to the pressure.

Oxygen, nitrogen, air and hydrogen are examples of gases which are almost *perfect* at moderate temperatures and pressures. The gases commonly used in refrigeration are not *perfect* under normal operating conditions, but, with the exception of carbon dioxide, their deviations from the gas law are not great at the temperatures and pressures normally prevailing in the low pressure side of a vapour compression refrigerating plant.

What is the difference between a gas and a vapour?

The gaseous substance in contact with the liquid from which it is formed is known as a vapour and it is still called a vapour if superheated to some extent. At still higher temperatures it is known as a gas but there is no sharp line of demarcation.

What is critical temperature?
Although in general a vapour may be liquefied by increasing its pressure up to the saturation value corresponding to the temperature, this is not so if the temperature is above a certain level. For any vapour this temperature above which no pressure, however high, will produce liquefaction, is called its *critical temperature*.

What is critical pressure?
The critical pressure of a vapour is the pressure required to liquefy it at the critical temperature and is the highest pressure on the temperature-pressure graph for saturated vapour. At temperatures above critical the pressure exerted by a vapour depends on the weight of it in a given space.

What is boiling point?
(1) When a liquid contained in a vessel having an opening to permit the exit of vapour is heated, its vapour pressure rises and eventually reaches that of the surrounding atmosphere. When this happens, the liquid boils and the temperature at which it does so is known as the boiling point for that particular pressure. It is therefore the temperature at which the saturation pressure of the vapour equals that of the atmosphere.
(2) When the liquid is in a closed vessel or system in contact only with its own vapour, the term *boiling point* is scarcely appropriate. It is preferable to speak of the *saturation temperature*, which is the temperature of the liquid surface corresponding to the pressure of the vapour in contact with it.

What happens to a gas which is allowed to expand in an engine cylinder?
Conversely to the compression of a gas, the work done by an expanding gas is produced at the expense of the heat contained in the gas, which therefore falls in temperature.

What is the triple point?
If a closed vessel containing only a pure liquid and its vapour is cooled, the pressure will fall with the temperature in

accordance with the pressure-temperature curve for saturated vapour (curve CP, Fig. 1).

Eventually, however, a temperature will be reached at which the liquid begins to freeze and any further reduction in temperature will cause the transformation of the whole of the liquid into solid. This temperature and its corresponding pressure are collectively known as the *triple point* for the particular substance. Only at this temperature and pressure can the substance exist simultaneously in solid, liquid and vapour forms.

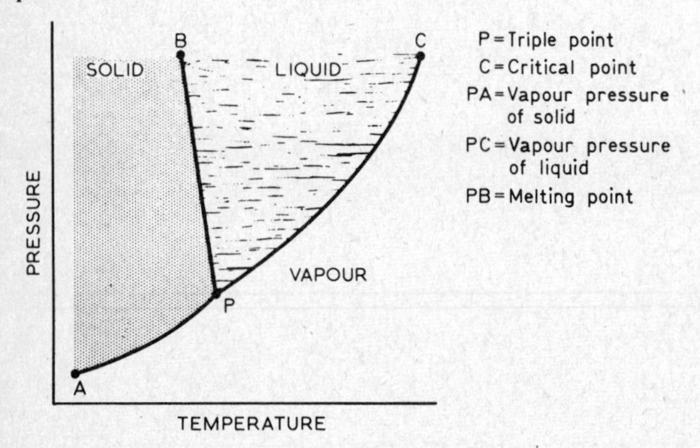

Fig. 1. Example of pressure-temperature relationship for solid, liquid and vapour.

What is the Joule-Thomson effect?

If a *perfect* gas is allowed to expand freely, without doing work (e.g. through a throttle valve) there is no change in its temperature. In practice, however, the temperature does change, if only slightly, because work is done by or against the inter-molecular forces. This change, known as the *Joule-Thomson effect*, may be a rise or fall; in the case of air and similar gases at high pressures and low temperatures it is a fall and this fact is made use of in the laboratory and elsewhere for the production of extremely low temperatures.

What are the principal ways in which heat is produced?

(1) The sun provides the earth with a continuous supply of heat.

(2) Many chemical reactions result in the formation of molecules which have a smaller store of heat energy than those from which they were produced, the difference being transferred to the surroundings which are thereby heated. Combustion is the most important case of such a chemical change. In human beings, animals, plants and even vegetables and fruits after gathering, a kind of slow combustion is always in progress, so that these also are sources of heat.

(3) Whenever two substances are in contact and one moves relatively to the other, the energy which has to be expended in producing the movement, that is to say in overcoming frictional resistance, appears as heat. Thus pumping water through a pipe, or circulating air through a duct by means of a fan, or driving a machine of any kind, are all processes in which the energy expended is converted into heat.

(4) The compression of a gas produces heat because the piston or other moving surface which pushes the molecules of gas forward "speeds them up" in doing so.

(5) Whenever an electric current flows, the electrical energy expended either appears as heat directly, as in the case of an electric heater or lamp, or part of it does so while the rest is converted into mechanical power and so ultimately into heat, as in the electric motor.

What is refrigeration?

Refrigeration is the transfer of heat from a substance to be cooled to somewhere else. As heat flows naturally from any body into any other colder body with which it is in contact, refrigeration is simple when a supply of some suitable colder substance is available. For example, fish can be cooled by packing ice around them.

Where a suitable colder substance is not available then one has to be produced, a complicated procedure involving the expenditure of energy; it is a process of this kind that is usually implied when the term *refrigeration* is used.

Nearly all refrigerating plants (but see Section 3) utilise the lowering of temperature which results from the controlled evaporation of a liquefied gas. When only small refrigerating effects are required they can be obtained by the direct application of electricity through a suitable thermocouple.

How does such a thermocouple operate?
In its simplest form a thermoelectric cooling device consists of pairs of semiconductor blocks connected in series and arranged as a sandwich, one face of which becomes hot and the other cold when a suitable direct current is applied. Therefore in effect, heat is taken from one side and discharged on the other, an ample heat sink being provided for its removal.

What is a heat sink?
A *heat sink* is a means for disposing of unwanted heat, usually by using it to increase the temperature of water, which is then run to waste.

What is a Frigistor?
Frigistor is simply the trade name given to a thermo-cooling device manufactured by one particular manufacturer.

What are the advantages of thermoelectric cooling?
(1) The device is small in size; a 15 A unit typically measuring less than 1 in square × ¼ in thick (25 mm × 6 mm).
(2) Because of their small size they are very suited to the small refrigerating effects often required in electronics, in instrumentation and for some medical purposes.
(3) They can produce very low temperatures: −130°F (−90°C), and lower in some circumstances.

Thermoelectric cooling is at the moment only viable, from the economic standpoint, in the smaller powers although this position will improve as their use increases enabling manufacturing costs to be reduced.

How is the evaporation of a liquid used to produce cold?
A liquefied gas is allowed to vaporise through a controllable

nozzle in such a way that the latent heat of vaporisation is taken from the substance to be cooled.

How is this accomplished?

In the simplest equipment the liquefied gas is allowed to discharge at a controlled rate through spray nozzles situated in the space to be cooled.

It follows that the gas used must be inexpensive (since it is discharged to waste), free from corrosive or toxic effects, non-inflammable and have as high a latent heat as possible.

Nitrogen is the gas most commonly used with this system which has advantages where fairly short period refrigeration is required, as in transport or overnight storage. The principle is also used in one form of in-line deep freeze equipment except that in this application means are usually provided for the recovery and reliquefaction of the gas used.

The use of liquid nitrogen for refrigeration is increasing, but by far the most common method of large scale and small scale refrigeration is one in which the gas used does not come into actual contact with the substance to be cooled and in which the gas is recompressed and again liquefied for re-use.

How can this be done?

By using one of two systems:

(1) The vapour compression system
(2) The absorption system.

(Refer also to Section 3.)

2

THE VAPOUR COMPRESSION SYSTEM

What are the fundamental facts on which the vapour compression system depends?

(1) The conversion of a substance from liquid to vapour form (evaporation) involves the absorption by the substance of a considerable quantity of heat (latent heat).

(2) For a given substance, the temperature at which evaporation or condensation (conversion from vapour to liquid form) takes place depends only on the pressure prevailing and can be varied by varying the pressure.

What are the essential parts of a vapour compression refrigerating system?

The evaporator, compressor, condenser and expansion valve or regulator (see Fig. 2).

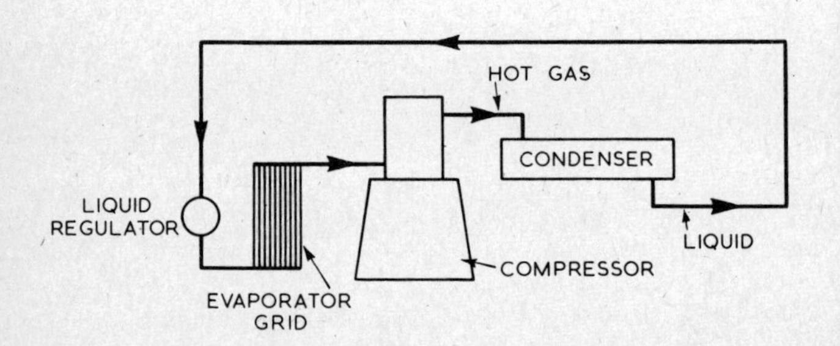

Fig. 2. Refrigerant circuit of vapour compression plant.

How does the evaporator function?
The evaporator is the most important part of the refrigerating plant, as it is here that the cooling effect is produced. It usually consists of some form of piping, immersed in water, brine, air or whatever medium it is desired to cool, and containing a suitable liquid (the refrigerant); or alternatively, the refrigerant may be contained in a casing and the medium to be cooled passed through pipes immersed in the refrigerant. In any case the function is the same; the temperature of the liquid is that corresponding to the pressure prevailing, and this temperature being lower than that of the adjacent medium, the refrigerant absorbs heat from the latter, and is thereby vaporised.

What is the function of the compressor?
By removing the refrigerant vapour as fast as it is formed, it maintains the required pressure in the evaporator, and it also raises the pressure of the refrigerant to a level sufficiently high to enable the condenser to perform its function.

What is the condenser for and how does it operate?
In the condenser, heat is transferred from the hot compressed refrigerant vapour to a cooling medium, usually water or air, and the refrigerant is thereby liquefied. This is done so that the same refrigerant may be returned to the evaporator and used again, the cycle of evaporation, compression and condensation being repeated indefinitely. For condensation to take place the compressor must be capable of raising the pressure of the refrigerant to such a level that the corresponding saturation temperature is higher than the temperature of the available cooling medium.

What is the function of the expansion valve (regulator)?
Its function is to control the rate at which the refrigerant circulates throughout the system; this must be such that the amount admitted to the evaporator in a given time is just as much as the heat absorbed by the latter can vaporise. If it does this correctly the regulator will necessarily also ensure that the most advantageous pressures and tempera-

tures for the particular installation, and for the external conditions prevailing, are maintained in the evaporator and condenser.

What is meant by evaporator duty?

The amount of heat which can be removed by the evaporator, i.e. the amount of refrigeration accomplished. It is measured in Btu/h or in tons refrigeration, or, where the metric system is in use, in kilojoules per hour (kJ/h) or kilocalories per hour (kcal/h). A kilocalorie is sometimes called a frigorie (f).

What is a ton refrigeration?

A *ton refrigeration* (TR) is that rate of removal of heat which would transform water at 32°F into ice at the same temperature at the rate of one ton in every 24 h. The basis is the U.S. ton of 2000 lb, and, taking the latent heat of fusion of ice as 144 Btu/lb, this means that

$$1 \text{ TR} = 2000 \times 144/24 = 12{,}000 \text{ Btu/h}.$$

What is meant by condenser duty?

The amount of heat which is transferred in a given time from the refrigerant to the cooling medium in the condenser.

What is indicated horsepower?

In the context we are considering, it is the rate at which work is usefully expended in the compressor, i.e. actually utilised in compressing the refrigerant vapour and expelling it from the compressor.

What is the relationship between evaporator duty, condenser duty, and indicated horsepower (i.h.p.)?

Since the refrigerant passes through a continuous recurrent cycle, the amount of heat which it receives in any given time added to the heat equivalent of the work done on it, must be exactly equal to the amount of heat removed from it. Hence: Condenser Duty = Evaporator Duty + Heat Equivalent of i.h.p. Any heat removed by jacket water must be added to the left hand side of the equation.

What are the principal factors affecting the amount of refrigerating duty (evaporation duty) obtained from a given plant using a particular refrigerant?
The most important is the *weight* of refrigerant circulated in a given time and this again depends on the density of the vapour drawn into the compressor, which in turn varies with the saturation temperature and pressure existing in the evaporator. The lower these are the lower will be the density of the vapour; hence, the lower the evaporation temperature the smaller will be the refrigerating duty obtained. Another relevant factor is the temperature of the liquid leaving the condenser; the higher this is, the lower will be the duty. This factor, however, is usually of minor importance except in the case of carbon dioxide where it has a great influence on the duty.

What other factors affect the evaporator duty?
Those mentioned above are inescapable even in a perfect plant; there are also certain sources of loss due to the imperfections of an actual plant. These all act by reducing the weight of vapour circulated by the compressor. The principal sources of loss are: clearance loss, throttling (loss of pressure) at the inlet and outlet of the compressor, heating of the vapour drawn into the cylinder by the warm walls and head of the latter and "slip" of the vapour past the piston during compression.

Under what circumstances are these losses greatest?
Without exception, they are increased by lowering evaporation temperatures and pressures and by raising condensation pressures and temperatures.

What is clearance loss and why is it important to keep the clearance volume of the compressor to a minimum?
The clearance volume of a compressor is that space between the end of the piston and the delivery valve which at the beginning of the stroke is full of compressed vapour. In a refrigerating compressor of conventional design, the suction valves are opened by the difference in the pressures existing

in the cylinder and in the suction pipe; consequently, the suction valve will not open to admit a fresh charge of refrigerant until the vapour in the clearance space has expanded sufficiently for its pressure to fall to that in the suction line. Thus, part of the effective stroke of the piston is lost. The proportionate loss depends on the clearance percentage (clearance volume expressed as a percentage of piston displacement) and also on the ratio of compression. Low clearance is therefore particularly important when low evaporation temperatures are required.

What is the adverse effect of heating of the vapour in the cylinder?
The density of the vapour is reduced, approximately in inverse ratio to the absolute temperature, therefore the weight of the charge drawn into the cylinder on each suction stroke is decreased.

Why are the losses due to piston slip and to cylinder heating even more serious than that due to clearance?
Because although clearance reduces duty and power (i.h.p.) in equal proportions, heating reduces duty but not power, and while piston slip results in some reduction of power, this is less than the reduction in duty.

What is a Mollier diagram?
A graphical representation of the properties of a fluid in which one of the co-ordinates is enthalpy (see Fig. 3). In the most usual and convenient form of the diagram, enthalpy is plotted as abscissa and pressure as ordinate. Besides the saturated vapour and liquid lines, the diagram carries families of lines for constant temperature, entropy and specific volume.

How is the Mollier diagram used?
Since evaporation and condensation occur at constant pressure, changes in the enthalpy of the refrigerant during these processes are equal to the heat which it removes from the medium cooled and which it gives up to the condensing water or air. Thus the evaporator and condenser duties for

unit weight of refrigerant can be read directly from the diagram.

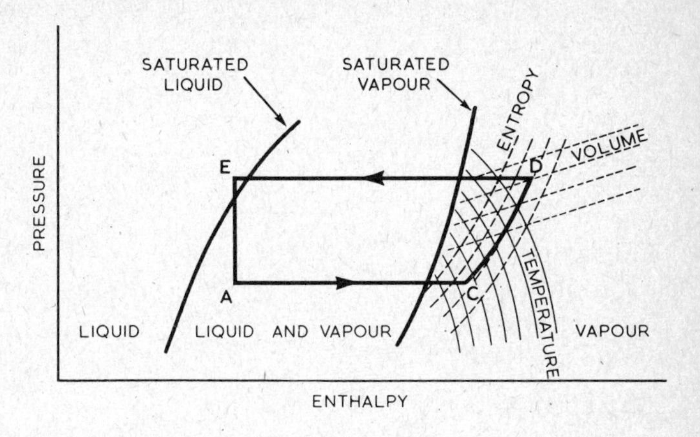

Fig. 3. Skeleton Mollier chart showing vapour compression refrigeration cycle.

What else, besides evaporator duty per unit weight, must be known before the refrigeration duty obtainable from a perfect compressor under given conditions may be calculated?

The swept volume of the compressor multiplied by the density of the refrigerant gives the weight of refrigerant pumped per hour by a perfect compressor, and this figure when multiplied by the evaporator duty per unit weight of refrigerant gives the amount of refrigeration obtainable from such a compressor.

How does the duty obtainable from an actual compressor compare with that calculated as described above?

The actual compressor will give the duty for an ideal compressor multiplied by the volumetric efficiency, which is always lower than 100% because of the losses previously described.

What volumetric efficiency may be expected?

The volumetric efficiency depends on the working conditions and on the design and size of the compressor. It varies widely and can be as low as 45% or as high as 95%.

How may the i.h.p. be calculated?

It may be obtained approximately from the Mollier chart once the evaporator duty is known. However, this method is not accurate, firstly because the weight of vapour compressed is a little greater than that circulating through the evaporator because some slips past the piston, and secondly because the heating of the vapour in the cylinder during the suction stroke and the early part of the compression stroke results in a somewhat greater i.h.p. than indicated on the Mollier chart.

What is the clearance volumetric efficiency?

It is the ratio of the weight of the refrigerant circulated by a compressor having no losses except that due to clearance, to the weight circulated by a perfect machine. It is difficult to estimate accurately because the extent to which the vapour becomes cooled during re-expansion is doubtful. An approximately correct result is given by the following formula, which is applicable to any of the common refrigerants:

$$C_{VE} = 100 - c\,(R^{0 \cdot 9} - 1), \text{ where}$$

C_{VE} = Clearance volumetric efficiency (%)
c = Volumetric clearance (%), and
R = Ratio of compression.

The diagram in Fig. 4 has been calculated on this basis.

How can the b.h.p. be found when the i.h.p. has been calculated?

The b.h.p. (brake horsepower or power which must be applied to the compressor shaft) is often arrived at by dividing the i.h.p. by the mechanical efficiency of the compressor. The latter is difficult to estimate, varying greatly with the operating conditions, and a better method is to add to the i.h.p. the estimated power lost in friction which although

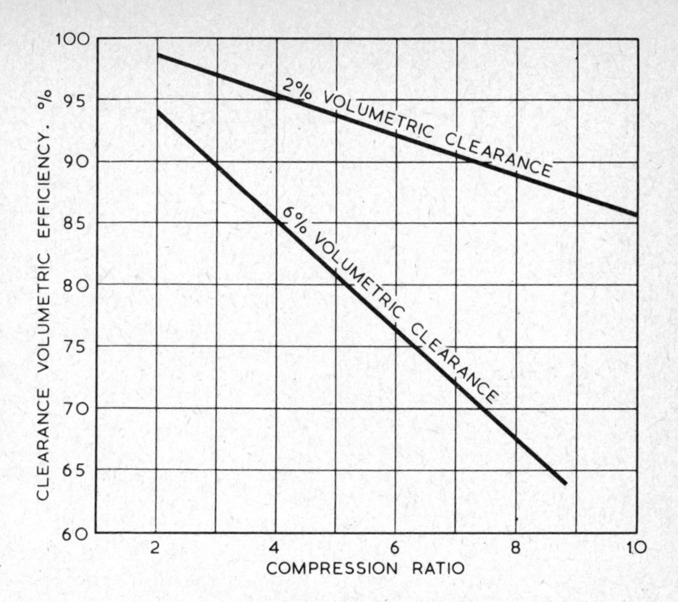

Fig. 4. Clearance volumetric efficiency.

not quite independent of operating conditions, does not vary greatly. Fig. 5 gives an idea of the friction hp, but the actual figure varies a lot with the design of the machine.

What is coefficient of performance (c.o.p.)?

The ratio of refrigerating effect to the heat equivalent of the i.h.p.

What is the highest value it can reach in a vapour compression plant?

The maximum possible value of an *ideal* cycle is given by:

$$T_1/(T_2 - T_1)$$

T_1 being the absolute evaporation temperature and
T_2 the absolute condensing temperature.

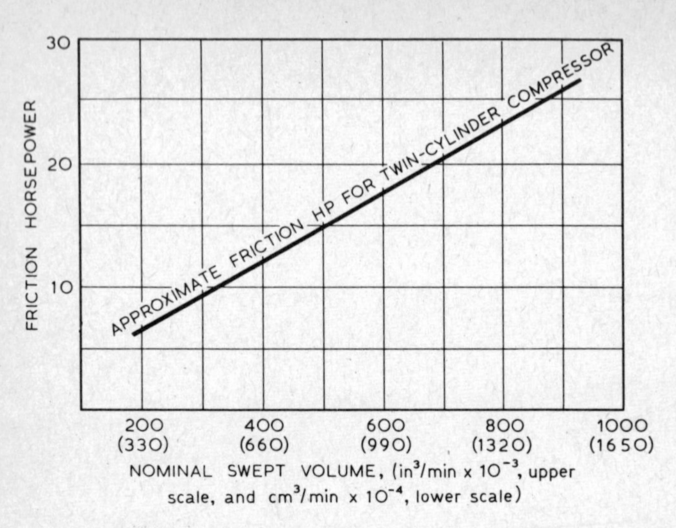

Fig. 5. Approximate friction horsepower for twin cylinder compressors.

How does the c.o.p. compare with the ideal value?

It is always less. Some refrigeration is lost in cooling the liquid passing through the expansion valve and the condenser process is not completely isothermal. The difference varies with conditions and the refrigerant used but is rarely less than 20%.

3

OTHER SYSTEMS

How does the absorption system operate?
The basic principle is the removal of heat from the substance to be cooled by evaporation of a suitable refrigerant—just as in the compression system; the difference lies in the means adopted for conveying the vapour produced to the condenser.

In a simple absorption system the vapour enters an *absorber* in which a weak solution dissolves the vapour thereby becoming stronger; heat is produced by this absorption and must be removed by cooling water or other means. The strong solution is then passed to a *generator* where it is heated, so driving off the refrigerant vapour, which passes on (together with some water vapour) to the condenser, while the weakened solution returns via a pressure reducing valve to the absorber. From condenser inlet to evaporator outlet the refrigerant undergoes changes similar to those which take place in the vapour compression system.

What working substance is used in an absorption plant?
A solution, the most usual one being ammonia in water.

What is the disadvantage of the absorption system?
The ratio of the refrigeration produced to the heat supplied is rather low, resulting in a higher running cost in comparison with the compression system.

Can anything be done to improve this ratio?
The economy of working may be improved by the addition of extra equipment. This may include an analyser in which the purity of the refrigerant vapour leaving the generator is

increased by contact with the incoming strong liquid. A rectifier which is connected in the circuit before the main condenser further purifies the refrigerant by selective condensation.

A heat exchanger can also be added between the absorber and the generator to reduce the heat required by the generator. These modifications are normally standard on domestic absorption types of cooling units.

What is an Electrolux cooling system?

Electrolux is the trade name of one manufacturer producing refrigerators of the absorption type that are charged with ammonia and water and pressurised with hydrogen. There is no expansion valve, the total pressure in all parts of the system being approximately the same. The source of heat may be a gas or oil burner, steam or an electric heater. The condenser is of finned tube construction cooled by the natural circulation of air.

Instead of the maintenance of a difference of pressure between the evaporator and condenser by the aid of an expansion valve, use is made of the fact that the pressure of the vapour of a liquid is independent of the presence or otherwise of a gas (non-condensible at the pressure prevailing) mixed with it.

The evaporator of the *Electrolux* unit contains hydrogen, a liquid seal preventing it from entering the condenser. Since the total pressure is the same in both vessels the pressure of the ammonia vapour, and hence the temperature of the liquid, is much lower in the evaporator.

The ammonia vapour formed in the evaporator, together with some hydrogen, passes to the absorber, placed at a lower level. The hydrogen with a much smaller proportion of ammonia vapour returns to the evaporator. The circulation is set up automatically by virtue of the lower density of the gas containing the greater proportion of hydrogen. Hydrogen gas is chosen because of its lightness and because it is practically insoluble in water. In the absorber, most of the ammonia content of the gas mixture is absorbed by a weak ammonia solution flowing from the generator. The absorber is cooled by air; the strong ammonia solution flows from it by gravity to the generator and the application of heat to the latter drives the weakened ammonia solution and refrigerant bubbles up a small diameter tube to the top of the generator. The weakened ammonia solution then returns by gravity to the absorber, whilst the refrigerant vapour rises to the condenser. Auxiliary devices used to improve the performance of the cooling unit are the analyser, rectifier and heat exchangers. Fig. 6 is a simplified diagram of the circuit, but the heat exchangers are omitted.

What is the largest size of absorption plant?
There is no practical limit to the size of plant although it is to be remembered that they are not economically competitive with electrically driven vapour-compression systems. Nevertheless, they are extensively used in small domestic cabinets where the difference in operating cost is negligible. They really come into their own in situations where gas (either town or bottled) or paraffin are the preferred means of providing heat. Units with capacities of up to 25 TR (75 kcal/h) are commercially available, usually for air conditioning work, and in the United States steam heated units in excess of 100 TR (300,000 kcal/h) are in operation.

What is the lowest temperature economically obtainable when using the vapour-compression or the absorption systems?
About −103°F (−75°C). It is possible to achieve lower temperatures by using two or more separate vapour-compression plants arranged in cascade. This is bulky and complicated, and specialist operating staff are necessary.

How is it possible to overcome these difficulties?
One method is to use a machine operating on the Stirling cycle (known also as "cold gas refrigeration"). Although this principle is by no means new, it is only during the last few years that it has come into its own, particularly where robust plants capable of temperatures down to −200°C (−328°F) are required.

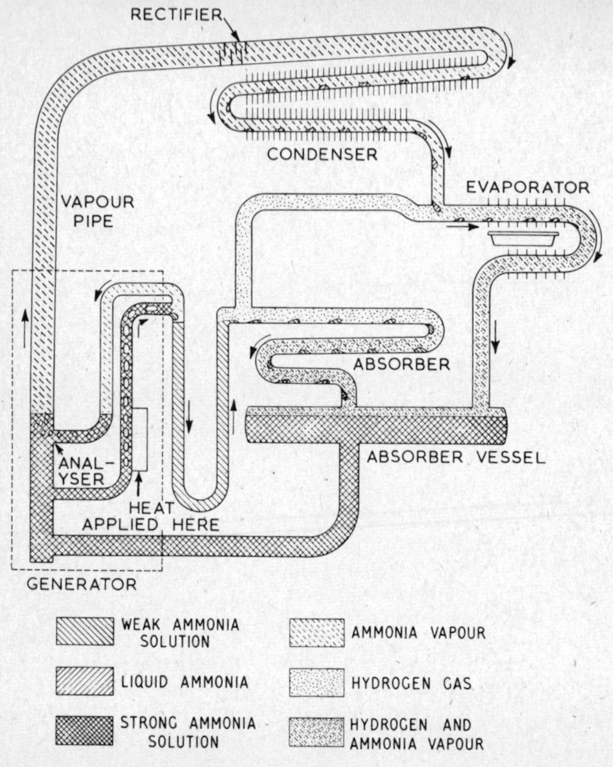

Fig. 6. The 'Electrolux' absorption cooling unit (courtesy Electrolux Ltd.).

What is the Stirling cycle?

The machine consists basically of two pistons, a heat exchanger, a heat sink and a sensible heat retainer, all constructed as one unit. The refrigerant remains as a gas throughout the cycle and in this respect is completely different to the systems already described. Hydrogen is commonly used as the refrigerant.

The sequence is as follows (refer to Fig. 7):

Stage (*a*) Piston 4 is at the top of its stroke and piston 5 is at the bottom, the space between being filled with refrigerant at low pressure.

Stage (*b*) Piston 5 has completed its upward stroke whilst piston 4 has remained still thus compressing the refrigerant into the space between them.

Stage (*c*) Piston 5 remains still whilst piston 4 descends to the top of piston 5. This pushes the hot refrigerant gas through the heat sink (3), through the sensible heat retainer (2) and the heat exchanger (1) into the space now available at the top of piston 4. In doing this the heat of compression is lost down the heat sink and the sensible heat is stored in the sensible heat retainer. Since the refrigerant then has a temperature equal to that of the substance being cooled no heat is given up or taken in through the heat exchanger (1).

Stage (*d*) Piston 4 and piston 5 descend together thus expanding the gas in the space above piston 4 and lowering its temperature. Piston 4 then returns to the top of its stroke

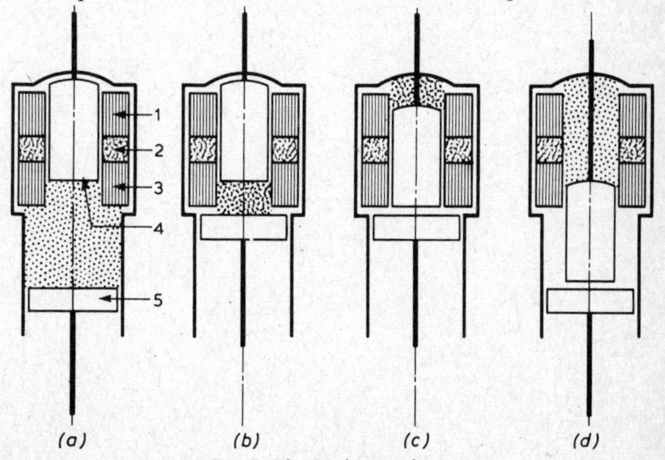

Fig. 7. The Stirling cycle.

1. Heat exchanger 2. Sensible heat retainer 3. Heat sink 4. Upper piston 5. Lower piston

whilst piston 5 remains still thus forcing the cold gas through the heat exchanger (1) where it takes up heat from the substance to be cooled, and also through the sensible heat retainer (2) where it again picks up the sensible heat left there on its upward journey. Because its temperature is equal to that of the heat sink no heat is lost there during the return journey when *Stage* (*a*) is again reached, and the cycle is repeated—usually between 1000 and 2000 times a minute.

4

PRIMARY REFRIGERANTS

What is a primary refrigerant?
A substance used as the working fluid in the vapour compression cycle, as distinct from a secondary refrigerant, which in some cases is used as an intermediate conveyor of heat between the substance to be cooled and the primary refrigerant. When the term 'refrigerant' is used without qualification primary refrigerant is understood.

What are the principal qualities required in a refrigerant for the vapour compression machine?
(1) High value of refrigerating duty per unit volume.
(2) Moderate working pressures.
(3) Chemical inertness and stability.
(4) Low cost.
(5) Non-corrosiveness.
(6) Non-toxicity.
(7) Non-inflammability.
(8) Moderate temperature after compression.

What are the refrigerants most commonly used today?
The principal refrigerants are shown in Table 1. *Freon*, *Arcton* and *Isceon* are the trade names used by specific manufacturers of the fluorocarbon refrigerants.

What are the advantages and disadvantages of ammonia?
Ammonia is used in a wide range of plant excepting those of quite small size. It has a high refrigerating duty per unit volume, and low cost, and it is practically non-corrosive to iron and steel. It requires moderate operating pressures but

Table 1. Refrigerants in general use

No.	*Chemical name*	*Formula*	*Pressure (lbf/in^2 at 5°F and 95°F and kgf/cm^2 at −15°C and 35°C)*		*Latent heat (Btu/lb at 40°F and kcal/kg at 4·4°C)*	*Boiling point at atmospheric pressure*	
						°F	*°C*
R.11	Trichloromono-fluoromethane	CCl_3F	23·95*	6·88	81·22	+74·7	+24·0
			610·0*	0·48	45·13		
R.12	Dichlorodi-fluoromethane	CCl_2F_2	11·79	108·0	65·71	−21·7	−30·0
			0·83	7·59	36·5		
R.22	Monochlorodi-fluoromethane	$CHClF_2$	28·33	183·6	87·39	−41·4	−41·0
			1·99	12·9	48·54		
R.40	Methyl Chloride	CH_3Cl	6·5	92·26	172·0	−10·6	−24·0
			0·46	6·5	95·56		
R.717	Ammonia	NH_3	19·57	181·1	536·0	−28·0	−33·0
			1·37	12·7	297·8		
R.744	Carbon dioxide	CO_2	317·2	1165·0	94·1	−109·0	−79·0
			22·3	82·0	52·3	sublimation	

NOTES:
1. R.11, R.12 and R.22 are also known as the *fluorocarbons*. There are others in this group not so generally used—such as R.13B1 (bromotrifluoromethane) and variants are frequently produced for special purposes. They are also known under trade names, such as *Freon*, *Arcton* and *Isceon*.
2. The SI unit of pressure is the newton per square metre (N/m^2). To convert lbf/in^2 to N/m^2 multiply by a factor of $6{\cdot}895 \times 10^3$, and to convert kgf/cm^2 to N/m^2 multiply by a factor of $9{\cdot}807 \times 10^4$.
3. * Measured in inHg vac. and cmHg vac.

is very toxic and, when mixed with air in certain proportions, is explosive.

What are the advantages and disadvantages of carbon dioxide?

It is non-toxic except in very high concentrations, inexpensive, non-inflammable and non-explosive. The refrigerating effect per unit volume is much higher than for any other refrigerant in common use but it requires high working pressures resulting in massive plant. The power required for a given refrigerating effect is higher than for any other refrigerant in common use. Nowadays its use is mainly confined to marine work.

What are the advantages and disadvantages of methyl chloride?

It requires moderate pressures, is moderate in cost and small leaks have no dangerous or unpleasant consequences. It is non-corrosive (unless moisture is present) to any of the usual engineering metals except aluminium and magnesium. It is not readily inflammable although it will burn, and it is toxic in considerable concentrations and over considerable periods of exposure.

What are the advantages and disadvantages of the fluorocarbon refrigerants?

They require moderate pressures and are non-inflammable and non-toxic but rather expensive. They are sensitive to moisture but despite this they are being increasingly used, especially in positions where their non-toxicity is of importance. That is, in situations where a refrigerant escape would constitute a risk to the life or health of persons in the close proximity of the refrigerator.

What is the difference between ammonia on the one hand and methyl chloride and the fluorocarbons on the other, with regard to behaviour with oil?

Liquid ammonia is only slightly miscible with lubricating oil and floats on the surface. It is therefore fairly easy to remove any oil which collects in an evaporator or elsewhere in the circuit, by draining. On the other hand, oil mixes freely with methyl chloride and with the fluorocarbons and therefore a special method must be used for separating any oil which passes into the evaporator, such as the distillation of the liquid refrigerant, or the circuit can be arranged so that oil is not left behind in the evaporator, the velocity of the vapour being sufficiently high to carry the oil back to the compressor in the form of a mist.

What is the effect of water in the refrigerant?

It may cause trouble through the formation of ice, particularly at the regulator. With methyl chloride and the fluorocarbons a corrosive acid is produced and under certain conditions an electrolytic transfer of copper from the piping to the cylinder walls etc. takes place, the lubricating oil acting as a "carrier".

How do the common refrigerants compare with regard to c.o.p.?

Under normal operating conditions they differ only slightly from one another in this respect, with the exception of carbon dioxide which has a much lower c.o.p. than the others.

How may refrigerant leaks be detected?

Ammonia by odour, by chemical test paper, or by hydrochloric acid or lighted sulphur stick, both of which produce dense white fumes in the presence of ammonia.

Methyl chloride by oil leakage or by soap bubbles after a soap solution has been applied to suspected parts.

The fluorocarbons by the use of a special halide detector torch which shows a change in flame colour in the presence of a leak. (It is possible to use the halide torch to detect leaks of methyl chloride but this should not be done unless the ventilation is very good.)

Refrigerant leaks may also be detected by automatic equipment operating on one of three methods:

(1) Two electric filaments form the balanced arms of a modified Wheatstone bridge circuit. One filament is made supersensitive by catalysts and reacts to the refrigerant by producing an increase in its running temperature thereby altering its resistance and causing an unbalance in the bridge, which can be used to sound an alarm.

(2) A difference in pressure within a detecting head results from the absorption of the refrigerant by special crystals. This pressure difference can be made to actuate a diaphragm and so sound an alarm.

(3) Depends upon the selective absorption of infra-red radiation by a gas. Every gas has a unique absorption spectrum and the amount of the absorption is a function of the gas concentration.

Detection is possible by these methods if the concentration is only 500 parts per million for ammonia, around 100 for the fluorocarbons and as low as 50 for carbon dioxide.

Chemical test paper may not detect very small ammonia leakage into water or salt brine; in these cases a more sensitive indicator is Nessler's solution, a few drops of which added to a sample of water or salt brine will turn yellow if ammonia be present.

If it is required to test a calcium chloride brine, it is necessary to dilute the sample with four times its own weight of water and then add sufficient concentrated sodium carbonate solution to precipitate all the calcium in solution. The sample should be filtered and then tested with Nessler's solution.

The degree of coloration, ranging from pale yellow through to deep brown, is an approximate index to the amount of ammonia present.

5

SECONDARY REFRIGERANTS

What is a secondary refrigerant?
Sometimes, instead of the evaporator being used to remove heat directly from whatever it is desired to cool, an intermediate fluid is used which is itself cooled by the evaporator. This may be called a *secondary refrigerant*.

What is its object?
Convenience, as in the case of ice making tanks of conventional design in which cans of water are frozen by immersion in cold brine; to facilitate the cooling of a number of different units by one compressor; to permit the easy adjustment of the temperature of the cooling surface; to provide a reserve of refrigeration for peak loads, and in cases where it is desired to keep the actual primary refrigerant well away from that which is being cooled. The last point is of particular importance where a leakage of the primary refrigerant would result in damage to the material being cooled, or in risk of injury to those working nearby.

What liquid is suitable for use as a secondary refrigerant?
In the great majority of cases a solution of calcium chloride in water, called *brine* by refrigeration engineers, is used. A solution of sodium chloride is also sometimes used.

What determines the freezing point of the brine?
The proportion of calcium chloride in the solution; this also determines its density.

What are the properties of calcium chloride brine?
These are listed in Table 2.

Table 2. Some properties of calcium chloride brine

Sp.gr. at 60°F (15·5°C)		1·08	1·11	1·15	1·20	1·25	1·28	1·30
Percentage of $CaCl_2$		9·26	12·53	16·75	21·79	26·59	29·35	31·1
Freezing point	°F	22·5	17·0	8·1	−6·4	−25·0	−41·0	−56·8
	°C	−5·3	−8·4	−13·3	−21·6	−31·7	−40·8	−49·0

How is density measured?
By using a hydrometer usually marked directly in specific gravity. Sometimes hydrometers marked on the Twaddell or Beaumé scales are used. Sp. gr. = (°Tw/200) + 1 and Sp.gr. = 145/(145 − °B).

How does temperature affect the measurement of density?
Density varies with temperature and the readings of a hydrometer do likewise.

What should be the brine density?
This depends upon the temperature of the brine (see Table 3).

Table 3. Recommended densities of calcium chloride brine

Lowest brine temperature °F	°C	*Specific gravity required*	*Amount of commercial grade $CaCl_2$ to make 1 gal* (lb)	*Amount of commercial grade $CaCl_2$ to make 1 litre* (kg)
+10	−12·1	1·225	4·25	2·1
0	−17·8	1·225	4·25	2·1
−10	−23·3	1·25	4·75	2·3
−20	−28·9	1·265	5·1	2·5
−30	−34·4	1·285	5·5	2·7
−40	−40	1·3	5·8	2·8

It is not advisable to use brines weaker than sp.gr. 1·225 because of the greater corrosive properties of the weaker brines.

Can anything be done to reduce the corrosive properties of calcium chloride brine?

Yes. The simplest method is to make sure the brine is slightly alkaline by the addition of caustic soda until a pH value of 8·5–9 is obtained.

What is the pH value?

To the chemist, it is the logarithm to the base 10 of the reciprocal of the concentration of hydrogen ions; in more everyday language, it is a measure of the acidity or alkalinity of a solution. Water, or a neutral solution, has a pH value of 7; greater values indicate alkalinity, lower values acidity.

How can the pH value be measured?

Electrical measuring instruments are available for the purpose.

What conditions tend to accelerate corrosion of metals by calcium chloride brine?

The presence of oxygen; when practicable air should be excluded from the brine circuit.

Can sodium chloride be used instead of calcium chloride?

Yes, it is sometimes used, but sodium chloride brine is extremely corrosive.

What is the eutectic point?

If the concentration of solid in a solution is increased the freezing point falls to a certain minimum temperature characteristic of the solution. This, the lowest freezing point obtainable, is called the *eutectic point* of the solution. If the concentration is increased beyond this critical value, then when the solution is cooled the solid will begin to separate out before the eutectic point is reached. This point for a calcium chloride brine is −59·8°F (−51°C) and for a sodium chloride brine −6°F (−21°C) and the lowest practical working temperature for calcium chloride brine is −40°F (−40°C).

What can be done if a secondary refrigerant is required to work at a temperature lower than − 40° ?
One of the following can be used:

Methanol with a freezing point around −140°F (−96°C).
Methylated spirit with a freezing point of −170°F (−112°C).
Trichloroethylene with a freezing point of −110°F (−79°C).
Ethylene glycol with a eutectic point of −40°F (−40°C).
Glycerol with a eutectic point of −60°F (−51°C).

The last two solutions are very viscous at the higher concentrations.

Does brine undergo any change during use?
Not if the brine is wholly within an enclosed system and therefore not exposed to the air; but if it is being used as a wash on plate coolers or for defrosting, it will absorb water and its freezing point will therefore rise.

What has to be done about this?
The brine solution has to be reconcentrated as often as necessary to maintain the required density.

How is reconcentration carried out?
There are several ways but the correct choice varies from case to case according to local conditions and to the degree and frequency of reconcentration found to be required. The simplest method is just to add more calcium chloride; this, however, can be messy, expensive in labour and costly in chemical. An alternative is to reconcentrate by boiling off the excess moisture. This can be done in large coppers fired with coke, oil or gas. Another method is to insert electric elements directly in the brine copper; this method can be costly in electricity but is easy and clean and lends itself to automatic working. Reconcentration has been successfully carried out by using the brine solution as a wash on atmospheric condensers but the condenser coils have a compara-

tively short life when this method is adopted. As a measure of the problem it may be of interest to note that a large cold store, say 1,000,000 ft^3 or 250,000 m^3 using brine washed air coolers requires the reconcentration of between 500 and 1000 gal (2250 and 4500 l) of brine every 24 h depending on the time of year.

6

REFRIGERATION COMPRESSORS

What are the principal types of compressor?
Reciprocating, rotary and centrifugal.

How many cylinders are fitted to a reciprocating compressor?
Up to as many as 16 arranged radially.

How are compressors usually driven?
Either belt driven or directly coupled to a prime mover. The prime mover is usually an electric motor but diesel engines, propane gas engines and hydraulic motors are sometimes used.

What is meant by the term "hermetic" applied to a compressor?
Hermetic compressors are those in which the motor and the compressor are manufactured as a single self-contained unit; the electric motor is in contact, therefore, with the refrigerant.

What is the semi-hermetic compressor?
In the *semi-hermetic* compressor the motor and compressor are a single unit but the motor is detachable from the compressor and therefore capable of field repair.

What is meant by the term "sealed unit" when applied to compressors?
This is a unit in which the compressor, usually rotary, and the driving motor are contained within a welded steel shell. Sealed units are incapable of field repair. This form of construction is usual for the smaller duties, as in domestic cabinets and similar applications.

How is compressor lubrication carried out?
Generally by splash for sizes up to approximately 10 hp (7·5 kW); some manufacturers, however, assist lubrication by means of a helical groove cut in the crankshaft which carries oil to the rubbing surfaces. Larger machines have forced lubrication using some form of oil pump.

What types of compressor valves are used?
Considerable variation exists; ring plate valves of various forms are most common in machines designed for moderate pressure refrigerants. Suction valves may be arranged in the top of the piston with the suction gas entering the machine via the crankcase or the cylinder wall, with the delivery valves in the compressor head. Alternatively both suction and delivery valves may be arranged in the compressor head. For very small compressors, valves consisting of thin flexible strips of metal are often used.

What is a safety head?
A *safety head* is one which is not fixed to the top of the cylinder but is held down by heavy springs. Its purpose is to allow the passage of liquid refrigerant, or oil, which it can do by lifting bodily whenever the pressure in the cylinder becomes abnormally high. All automatically controlled machines, unless of the very smallest size, should be provided with safety heads because of the possibility of considerable quantities of liquid being pumped, particularly when the compressor starts. If a safety head is not provided, the delivery valves may be quite inadequate to pass the liquid and the machine may be damaged if not wrecked.

What is the value of cylinder cooling and how is it effected?
The objects of cooling the cylinder are: (*a*) to avoid excessively high delivery temperatures, and (*b*) to reduce the i.h.p. Machines are often fitted with water jackets around the upper part of the cylinders or may have fins cast integrally with the cylinders to assist in dissipating the heat of compression to the air.

How may swept volume be varied?
(1) By the use of a variable speed electric motor.
(2) By the introduction of a throttling valve in the suction line, either hand or automatically operated, to maintain a constant pressure in the evaporator. It is a simple method of reducing the duty to a moderate extent but should not be used where large reductions are required because of the very low suction pressures which result.
(3) By providing a variable clearance pocket in the compressor.
(4) By providing arrangements for putting one or more cylinders out of action—such as holding the suction valve permanently open or by by-passing the cylinder (see Fig. 8).

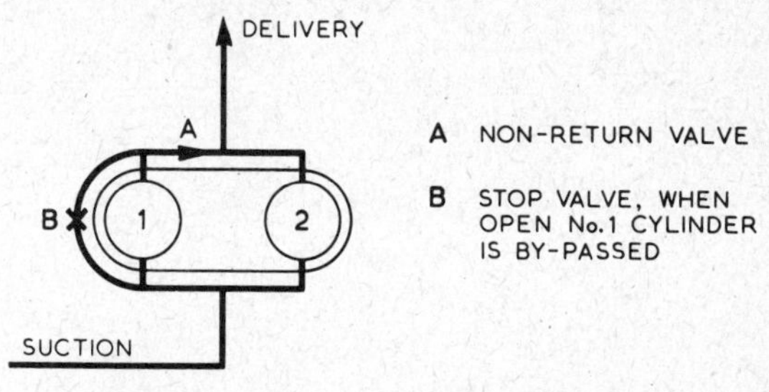

Fig. 8. Capacity reduction by by-passing a cylinder.

Sometimes a port is provided in the cylinder wall at the half stroke or some other suitable position; this can be put in communication with the suction side by opening a stop-valve so that during the first part of the compression stroke the refrigerant is simply returned to the suction main. This port in the cylinder wall is sometimes used as a means of obtaining two different suction pressures from the same machine which can then operate evaporators at two different temperatures.

What is a rotary compressor?
A standard type consists of a cylindrical casing containing

a shaft whose axis is eccentric to that of the cylinder; the shaft carries a rotor having radial slots in which slide blades pressed against the casing by their inertia or with the aid of springs. The ratio of compression obtainable depends on the relative sizes and positions of the suction and discharge ports. A non-return valve is fitted on the delivery side. Rotary compressors are compact and run without pulsation or vibration, and at high speeds. They can have high volumetric efficiencies. Rotary compressors are used extensively in household refrigerators and similar applications and also in very much larger sizes as booster compressors.

What is a centrifugal compressor?

It is similar in principle to a centrifugal pump, the necessary increase in the pressure of the refrigerant vapour being obtained by imparting a high velocity to it by the rotation of an impeller. The "velocity energy" thus obtained being converted into "pressure energy" by gradual reduction of the velocity through diffusers.

The density of the vapour being relatively low in comparison to the liquids dealt with by centrifugal pumps, the rise in pressure per stage is not great; the heavier refrigerant vapours must be used; several stages are necessary and even so the utility of this type of compressor is confined to applications involving high evaporator pressures, such as air conditioning work for example. The centrifugal compressor is only suitable where large volumes of vapour have to be circulated, because of the high impeller velocities which are essential to provide the required rise in pressure. It should also be noted that since the pressure rise depends on the speed of rotation, it is not independent of the volume being circulated and capacity control is therefore not obtainable by varying the speed of the compressor.

7

EVAPORATORS

How fast is heat taken in by an evaporator?
Heat taken in = Surface × Coefficient of heat transfer × temperature difference.

What is the coefficient of heat transfer?
The answer can be made clearer by not thinking in terms of the heat transfer coefficient (U) but of its reciprocal which is the resistance to heat transfer and is analogous to electrical resistance. Then the resistance to the transfer of heat through a material is the sum of three separate resistances: film resistance on one side of the material, the resistance of the material itself and the film resistance of the other side of the material. Except in the case of insulating materials the second of these resistances is so small relative to the other two that it may be neglected. A further resistance due to the presence of a film of scale or dirt must be allowed for in a practical design.

What is the resistance on the refrigerant side of an evaporator?
The resistance to the transfer of heat from an evaporating refrigerant to a pipe varies widely and depends on many factors; it can be 100–500 ft^2 h °F/Btu (500–2500 m^2 h °C/kcal). It just depends on the detailed design and the refrigerant in use.

On what does the resistance on the cooled medium side depend?
The principal factors are the following: the nature of the medium being cooled, e.g. water, brine, air (in order of increasing resistance); the velocity of flow of the fluid (the

higher the velocity, the lower the resistance); the diameter of the pipe; whether the fluid is inside or outside; the direction of flow; and, if the fluid is outside, whether the pipes are in line or staggered. (Small pipes provide lower resistance to heat flow; fluid flowing across pipes meets with much lower resistance than fluid flowing inside them and staggered pipes have a rather lower resistance than pipes in line.)

How are actual transfer coefficients determined?
By practical test. The cost of an evaporator for a given duty depends upon the surface area provided. The smaller this surface is the more competitive the design. All manufacturers publish tables showing the duties obtainable from their own particular designs and these are reliable provided the conditions specified for the duties stated are carefully noted and complied with.

Since the temperature of the medium being cooled changes as it passes through the evaporator, how may the effective temperature difference be calculated?
Provided that the specific heat of the medium cooled is constant—usually approximately true for moderate reductions in temperature—the temperature difference is the logarithmic mean of $T_1 - T$ and $T_2 - T$ where T_1 and T_2 are the inlet and outlet temperatures respectively of the fluid cooled and T is the evaporation temperature.

What is the effect of pressure drop through the evaporator?
Wherever boiling liquid exists in the evaporator the temperature corresponds to the pressure. If, therefore, the resistance to flow is such as to cause an appreciable drop in pressure the temperature of the refrigerant will fall as it progresses through the evaporator. For a given duty to be obtained from a given compressor, the evaporator gauge (pressure and temperature) at the suction must be correct; this being so the effect of pressure drop is that it would obviously mean that the temperature of evaporation is

higher than it would otherwise be, and the temperature difference for heat transfer less—requiring more evaporator surface.

What is the maximum allowable pressure drop?

There is no definite limit to the pressure drop permissible but 3°–5°F (2°–3°C) is as much as should be allowed.*

Into what two classes may evaporators be divided, with reference to circulation of refrigerant?

Flooded and *non-flooded*.

What is the distinction between the two classes?

In a *non-flooded* evaporator, there is a continuous circulation of refrigerant from the inlet (where it is mainly in liquid form) to the outlet (vapour) and the speed of this movement corresponds exactly to the rate of pumping of the compressor and (if regulation is correct) to the rate of evaporation. The quality of the refrigerant (proportion of vapour) increases steadily as it moves through the evaporator.

In the *flooded* evaporator, on the other hand, the liquid refrigerant circulates at a rate independent of that at which it is evaporated, and many times greater than the latter; consequently, the proportion of surface in contact with liquid at any moment is much greater and is more nearly uniform throughout the evaporator.

Why is the non-flooded type of evaporator generally used in small plants?

Its construction, particularly in small sizes, is much simpler; when the total cost of the plant has to be small, the more elaborate construction of a flooded evaporator is not warranted. Again, small plants generally use one of the fluorocarbon refrigerants; the lubricating oil mixes with these liquids and the uni-directional flow of refrigerant facilitates the return of the oil to the compressor.

* Pressure drop is frequently expressed in degrees of temperature denoting the corresponding temperature drop.

Why is the flooded type of evaporator very widely used in large plants?
The fact that the refrigerant side is thoroughly wetted ensures a high coefficient of heat transfer, which in large units results in sufficient saving in surface to more than offset the more elaborate construction. Another factor is that $1\frac{1}{2}$ in (40 mm) nominal bore piping being about the largest size practicable for forming into coils and grids, the number of circuits necessary to avoid excessive pressure drop becomes very large in a large plant and with a non-flooded system this would involve complicated distributive devices or the use of a large number of regulators.

How is a non-flooded evaporator made?
When used for cooling another liquid such an evaporator usually consists of a tank containing the liquid to be cooled and in which is immersed the evaporating coils—usually known as *grids*. There may be more than one pipe circuit. The grids may be directly attached to the walls of a chamber to be cooled—in which case they are known as "direct expansion circuits"—and in these cases are often of great length and in many separate circuits.

What is a shell and tube evaporator?
This consists of a cylindrical casing, usually horizontal, containing a number of tubes fitted into end plates closing the ends of the casing. Over these are fitted covers. The refrigerant is contained within the casing whilst the liquid to be cooled flows through the tubes. Generally the end covers are sub-divided so as to cause the liquid to be cooled to flow backwards and forwards several times; thus if there are 50 tubes these might consist of 10 passes each having 5 tubes in parallel. A vapour dome or header is usually fitted astride the casing to separate liquid entrained with the suction vapour.

What is a Baudelot evaporator?
An open type of cooler in which the liquid to be cooled flows from distributing troughs or headers over a cooling surface

consisting of sets of grids, or a pair of stamped corrugated metal sheets forming channels. It may be flooded or non-flooded. If made of pipes, these should be close together to avoid splashing of the liquid.

What is an accumulator?

An *accumulator* is a steel shell partly filled with liquid refrigerant. The space above the liquid is maintained by the compressor at a pressure corresponding to the required refrigerant temperature. The refrigerant liquid is pumped from the accumulator through the cooling coils of an air or brine cooler and returned (together with a certain amount of vapour) to the accumulator.

What overall heat transfer coefficient is obtainable from a flooded shell and tube evaporator?

This depends upon the detailed design of the evaporator, the velocity of the liquid being cooled and the mean temperature of the liquid being cooled. With a velocity through the tubes of the order of 5–6 ft/s (1·8 m/s) a heat transfer coefficient somewhere between 60 and 100 Btu h^{-1} ft^{-2} $°F^{-1}$ (300–500 kcal h^{-1} m^{-2} $°C^{-1}$) is usually realised; but as has been stated, a lot depends on the detailed design of the cooler and on its cleanliness.

8

AIR COOLERS

What forms may an air cooler take?
An air cooler consists essentially of cold surfaces over which air is passed by a fan.

The cold surface may consist of steel sheets cooled by drenching them with cold brine, or of tubes kept cool by the refrigerant passed through them. These tubes may have gills moulded to them to increase their surface, both inside the tube and outside the tube. Some air coolers contain mixtures of flat plate and pipe grids but present day practice tends towards the use of tubing gilled on the outside only.

What is the purpose of adding gills to tubing?
To increase the heat transfer area and to compensate for the fact that the heat transfer coefficient inside the tube, being wet, is greater than that on the outside of the tube normally in the air flow. For maximum effect the amount of gilling should be such that the heat transfer per foot run of tube is the same inside and outside.

What is the best spacing of tubes?
This depends on whether one requires maximum heat transfer, minimum resistance to air flow, or a compromise between the two. It depends also to some extent on manufacturing possibilities (see Fig. 9).

How is gilled tubing made?
In some cases, particularly for small coolers, flat rectangular copper plates are used and are threaded on copper pipes which are expanded to make firm contact with the plates. Bends are brazed or sweated onto the pipes and the whole

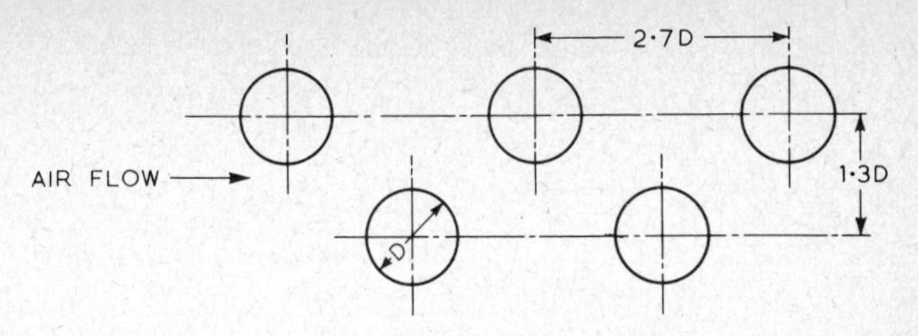

Fig. 9. Optimum spacing for pipes in a cross-grid cooler.

lot is usually tinned. Another form is made by winding a steel or copper ribbon on a pipe of the same material, the part of the ribbon towards the tube becoming crinkled in the process. The gilled tubing may be tinned after manufacture if it is made of copper or galvanised if it is made of steel, to ensure good thermal contact between gill and tube.

Another method is to make the gills an interference fit on the pipe, assemble the necessary gills for a length of tube with temporary spacers between them, and then force the tube through the whole assembly by hydraulic pressure. The spacers, in two parts, are then removed together with the former which held it all in place. This method is much used where two or more pipes are required to share a single gill (see Fig. 10).

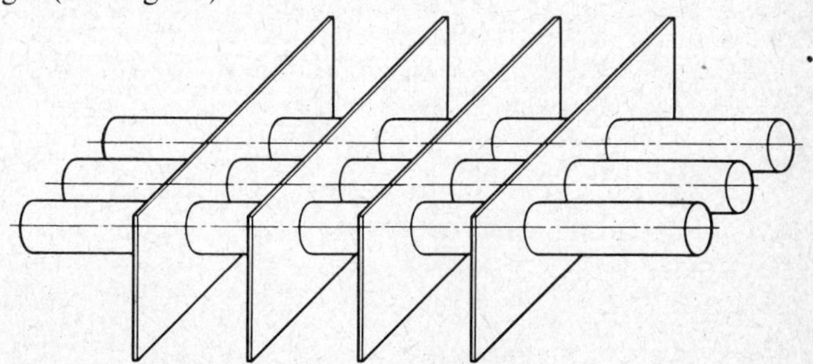

Fig. 10. Sketch showing three pipes sharing a single gill—the centre pipe of the three can be used for defrosting purposes only.

What is frosting?

The surface of an air cooler is usually colder than the air passing through it; the air can be cooled below dew point and when this takes place there is a throw-out of moisture which appears upon the surface of the air cooler as snow or ice.

Does this frosting matter?

Yes, for two reasons: it seriously reduces the heat transfer coefficient of the cooler surfaces, and, if allowed to accumulate, it will fill all the air spaces with solid ice or hard snow and thus shut off the greater part of the air supply.

How can this frosting-up be prevented?

It can be prevented if the air cooler is so designed that the air passing through never reaches its dew point; this method is usually only possible when the temperature being maintained is more than 40°F (5°C). Alternatively, the surfaces of the cooler can be brine washed. Although this will prevent the formation of snow on the cooler surfaces dilution of the brine takes place with the result that the action of defrosting is replaced by one of reconcentration. Usually the formation of snow and ice is accepted and defrosting undertaken regularly.

How often should a cooler be defrosted?

This depends very much on the detailed design of the cooler and on its working conditions. With the larger commercial applications defrosting is carried out at least daily and sometimes, particularly when the operation is automatic, more frequently than this.

How does a brine washed cooler prevent frosting up?

Because the brine takes the moisture "thrown out" of the air into its solution therefore diluting it, as stated previously.

Are there any advantages in brine washing coolers?

Yes, the heat transfer coefficient of the cooler surfaces is more than doubled.

How is the brine reconcentrated?
This has been dealt with on page 41.

How may an air cooler be defrosted?
There are several methods:
(1) Washing copiously with water.
(2) Reversing the refrigerant cycle thereby passing hot gas through the evaporator coils.
(3) Passing a hot liquid through special tubes installed ajacent to the evaporator tubes.
(4) By placing electric heating elements next to the evaporator tubes.
(5) By blowing warm air over the evaporator.

What are the advantages of each of these methods?
Drenching the cold surfaces with cold water is straightforward and simple but some care is necessary to make sure that there is sufficient water to remove the snow or ice without its own temperature being lowered too much. There must also be really adequate drainage arrangements since any blockage will result in the drenching water spilling out onto the floor where it may freeze—a very real problem. The water required is considerable and the Water Authorities are becoming increasingly more reluctant to supply water for this purpose.

Reversing the refrigerant cycle is often used but usually requires skilled attention, and of course, there must be an adequate load for the compressor while defrosting by this method is being carried out.

The use of a hot liquid, which is usually glycol or brine, running through special tubes is very frequently adopted (see Fig. 10). Sometimes there is one defrost tube to two evaporator tubes and sometimes one defrost tube to each evaporator tube. Special piping circuits and pumps are required as well as calorifiers, which are heated by hot water so as to reheat the defrost liquid once it has passed through the defrost tubes. Care is required in interconnecting the various parts of the refrigerating plant to make sure that defrosting is not accidentally switched on at the same time as refrigeration is being applied.

The use of electric elements placed adjacent to the evaporator tubes is becoming more and more popular, particularly in the smaller sizes of air cooler. The electric power required is considerable but the system is very suitable for automatic operation—usually with the aid of a time clock.

Defrosting by blowing comparatively warm air over the coils is normally only possible where air temperatures in excess of 40°F (5°C) are being maintained. The operation is simple; refrigeration is switched off but the fan is allowed to continue to run.

What happens to the ice and snow melted by defrosting?
Most of it falls off in lumps or drips into a catchment tray placed under the cooler. Where defrosting is carried out from inside the snow deposit—as in methods (2) and (3)—there is usually more ice than water. With the other methods it is mostly water with only a few ice lumps.

Does this ice and water re-freeze in the catchment tray?
It will do so unless suitable precautions are taken. The trays must be insulated and heated—usually by electric elements placed under or in them. The drain must also be insulated and, where it passes through refrigerated space, also wrapped with heating tape. There should not be any "U" traps in the drain, but should any consideration make their use really necessary then special tailored heated bonnets or jackets are necessary around them.

Does the heat used for defrosting pass into the cold space?
If the amount of heat applied is properly controlled all of it will be used to provide the latent heat required to melt the ice or snow. Any excess will pass into the cold space. In this connection it is useful to remember that the heat applied to the catchment trays usually needs to be switched on a good deal longer than the heat applied to the evaporating tubes themselves and for this reason independent control is necessary.

Can hot gas defrosting be applied to the absorption plant? Yes, such a device is fitted to the larger domestic refrigerators manufactured by Electrolux Ltd. The defrosting action is brought about by diverting hot gas from the generator to the inside of the lower cooling section of the evaporator at periodic intervals. The operation is entirely automatic and its frequency is determined by an automatic siphon arrangement in the generator system which periodically empties an associated siphon chamber, allowing hot gas to pass through the chamber and hence along a by-pass pipe to that section of the evaporator which is iced up.

9

CONDENSERS

What types of condenser are in common use?

Water cooled: shell and tube; double pipe; shell and coil.
Air cooled: forced draught.
Evaporative: forced draught.
Atmospheric: i.e. evaporative with a natural circulation of air.

How is the shell and tube condenser made?

It consists of a cylindrical shell, usually horizontal, with a number of tubes fitted into end plates, and covers arranged to give a suitable number of passes for a water velocity generally of 3–5 ft/s (1–1·5 m/s). Very similar, indeed, to the shell and tube evaporator except that no dome is required. Vapour enters at the top and liquid is drawn off from the bottom. This type of condenser is widely used on all classes of plant above approximately 3 hp (2·25 kW).

What coefficient of heat transfer is obtainable?

This depends upon the refrigerant in use, the water velocity and the detailed design of the condenser and varies considerably (between 100–200 Btu h^{-1} ft^{-2} $°F^{-1}$ (500–1000 kcal h^{-1} m^{-2} $°C^{-1}$) with a reasonably clean condenser; less if the tubes are very scaled).

What is the resistance to water flow through a shell and tube condenser?

This, again, depends very much on the details of the design particularly of the end covers, and is most difficult to predetermine except on the basis of experience. Manufacturers quoted figures are based on practical tests and can be relied upon.

How can the quantity of water required by a condenser be calculated?
Flow = Condenser duty/10 × Temperature rise. The units are gal/h, Btu/h and °F respectively. A reasonable temperature rise is 15°F (8°C). The flow in litres per hour is given by: Flow = Condenser duty ÷ Temperature rise. Condenser duty is in kcal/h and temperature rise in °C.

What temperature difference should be used in calculating the condenser surface?
The condenser has two important duties, and sometimes a third minor one. It has to perform the removal of superheat from the refrigerant vapour, the condensation of the vapour, and in some cases, the cooling of the liquid to a few degrees below the temperature of condensation. The temperature differences and the coefficients of heat transfer for these duties are all different, so that tests under varying conditions do not give consistent heat transfer (U) values when based on a logarithm of mean difference of temperature condenser gauge water inlet and outlet. This, however, is the normal basis and is satisfactory for all practical purposes.

How would one describe the double-pipe condenser?
It consists of concentric tubes made up to form grids. Usually water flows through the inner tube. It has been largely replaced by the more compact shell and tube type.

What is a shell and coil condenser?
This is a useful form of condenser for small duties; its heat transfer coefficient is quite good and its construction simple. It consists of a water coil in a cylindrical shell, usually vertical, in which the refrigerant is condensed.

How is an air cooled condenser made?
In very small sizes it sometimes consists of a simple coil of pipe, cooling being effected by natural convection and radiation. More generally it is made from stacks of gilled tubing

with headers at the top and bottom, over which air is forced by a fan.

What temperature difference should be used as a basis for design?

The actual temperature difference will vary as working conditions vary, but a reasonable basis for design would be to take the condenser gauge 20°F (11°C) higher than the air flowing onto the condenser.

What is an atmospheric condenser?

An *atmospheric condenser* consists of piping arranged in the form of grids with water circulated from a tray underneath by means of a pump. The water is distributed at the top by a trough or by sprays from where it runs down over the pipes. The water, being heated by the pipes and cooled by its own evaporation into the surrounding air, reaches a steady temperature which depends upon the rate of heat flow on the one hand and the rate of evaporation on the other. The latter is dependent on the wet bulb temperature of the air so that the condenser is more effective the drier the air. This type of condenser must be exposed to the external atmosphere and is often erected on the roof of a building. The only water required is that necessary to make up for evaporation plus any loss due to splash or drift—say 1 gal for every 3000 Btu (750 kcal) absorbed by the condenser—more if exposed to high winds.

What is an evaporative condenser?

The *evaporative condenser* is similar in principle to the *atmospheric* type but it is provided with a fan for forced circulation of air. It is used in cases where scarcity or high cost of water make the shell and tube condenser undesirable. A similar result is obtainable by using a shell and tube condenser and causing the water to cool itself by circulating it over a cooling tower.

10

REGULATION

What is the function of a regulator?
Usually called the expansion valve, the function of the regulator is to control the rate at which refrigerant is admitted to an evaporator. All other functions are incidental.

What are the types in common use?
The hand expansion valve, the constant pressure valve (sometimes called automatic), the high pressure float valve, the low pressure float valve and the capillary tube.

What is a hand regulator?
This consists simply of a valve similar to a hand stop valve but with a finely tapered spindle. It is set to give the required rate of admission by turning a hand wheel. Until recent years it was the only kind of regulator in general use but is now rarely seen.

Supposing a hand regulator is set so that only part of the evaporator is used as such, the remainder superheating the vapour, what happens if the regulator is opened further?
The rate of flow of refrigerant into the evaporator is immediately increased, but the rate of flow of heat remains the same at first, and so does the volume of vapour drawn off by the compressor. Consequently, the volume of evaporator occupied by liquid increases and so does the wetted surface. Thus the heat intake increases; the volume of vapour produced becomes greater, and since the compressor cannot increase its pumping capacity, the vapour becomes compressed into a smaller space and its pressure rises. The temperature of

the boiling liquid rises correspondingly, reducing the temperature difference for heat transfer, and the heat intake therefore decreases again.

Thus a new condition of equilibrium is eventually attained, with a higher evaporating temperature and pressure than before, greater heat transmission, more of the evaporator in use as such, and a lower temperature difference. The establishment of this new condition is not instant, because the changes in the quantity and temperature of liquid in the evaporator take time. Assuming a sufficient reserve of liquid to be available in the liquid receiver, the process can continue until a point is reached at which liquid begins to be carried with the vapour entering the compressor, after which the duty falls off sharply.

How does the constant pressure regulator operate?

The flow through a hand regulator and the evaporator pressure obtained, remain constant only so long as the condenser pressure and the temperature of the material cooled are unchanged. Where it is desired to maintain an evaporator pressure constant under all circumstances, a constant pressure type of regulator, which is self-adjusting, may be used. In a typical regulator of this type the valve is balanced between the forces exerted by a spring, together with the atmospheric pressure on the one side, and the evaporator pressure acting through the intermediary of a diaphragm and usually assisted by a second spring, on the other. Any rise in evaporating pressure causes the diaphragm to close the valve slightly against the spring force, and conversely, a fall in evaporator pressure results in the valve opening slightly. In each case the movement continues until the pressure is restored to its previous value.

The valve is provided with an adjustment for the spring compression, by means of which any desired evaporating pressure can be maintained.

What is a thermostatic regulator?

A *thermostatic regulator* (see Fig. 11) is designed to adjust the flow of refrigerant in such a way that as much of the evaporator as possible is always in use, as such, and that no liquid refrigerant reaches the compressor, i.e. it regulates the plant just as a competent operator would with a hand regulator in order to obtain the maximum duty from it under all circumstances. This result is obtained by maintaining an approximately constant superheat, the temperature of the refrigerant in the suction line to the compressor being a fixed number of degrees (usually about 15°F (8°C)) above the evaporating temperature. In construction it resembles the constant pressure regulator but incorporates an additional device, the thermal element, which provides a force corresponding to the temperature of the refrigerant in the suction line.

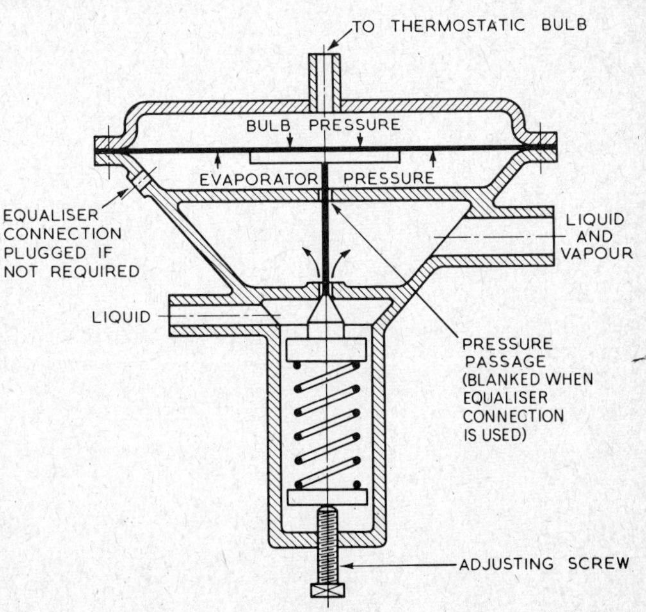

Fig. 11. Functional diagram of thermostatic regulator.

In its most usual form this consists of a small bulb or phial, usually containing some liquid refrigerant fastened in good thermal contact with the suction line, and connected by a small bore tube to a diaphragm or bellows in the regulator. The valve is balanced between a force due to the evaporator pressure on one side and a force due to the pressure in the phial (which itself depends on the suction temperature) on the other. Springs are fitted to balance these forces, for stability and to allow for adjustment. Therefore if the difference between suction temperature and evaporator gauge (i.e. the suction superheat) increases, the valve opens a little, admitting more refrigerant and restoring the original conditions. If the superheat decreases, the valve closes a little so that less refrigerant is admitted to the evaporator and the point in the evaporator at which superheating begins recedes.

It may be noted however that it is actually the difference between the *pressure* in the bulb and that in the evaporator which is held constant; the extent to which the constant temperature difference is obtained when the evaporation temperature varies depends mainly on the temperature-vapour-pressure characteristic of the liquid in the bulb. This regulator has a very wide field of application and its use is particularly desirable where the temperature of the medium cooled varies over a wide range, as in cooling a bulk of fluid from a high to a low temperature. It is also advantageous, even with fixed evaporator conditions, when an air-cooled condenser is used, as it compensates for the variation in the weight of liquid entering the evaporator with varying condensation temperatures.

Is a thermostatic regulator suitable for controlling flooded circuits?

No. In such circuits it is important to control the liquid level closely, and thermostatic regulation is not constant enough for this; it permits a fluctuation of suction superheat and of liquid level about the mean.

What is an equaliser connection?
A thermostatic regulator valve maintains an approximately constant difference between the two refrigerant pressures which actuate it; these are the pressure of the vapour in the sensitive bulb and, normally, that of the vapour immediately after expansion. If there is a considerable drop in pressure through the evaporator, the pressure at the regulator will be considerably higher, i.e. the difference between the suction line temperature and the saturation temperature corresponding to the suction pressure, will be abnormally high. To avoid the abnormally high superheat in cases such as this, the passage in the regulator through which the expansion pressure normally acts on the valve is blanked off, and instead a small tube is connected from the regulator to the compressor suction; thus the pressure actuating the valve is the suction and not the expansion pressure.

When must an equaliser connection be fitted?
This depends on the amount of superheat adjustment which the design of the regulator permits. Generally, an equaliser connection is necessary when the total pressure drop from regulator to bulb position exceeds the equivalent of 6°F (3°C). This will always be the case if the evaporator is divided into several circuits to which refrigerant is distributed through orifices.

What is a high pressure float?
A simple float-operated needle valve, the valve controlling an orifice at the outlet from the float chamber, so that the latter is under condenser pressure. All the liquid refrigerant entering the chamber leaves at the same rate as it enters since the level in the chamber is held constant.

Why must the high pressure float chamber be vented?
Since this type of float drains all the liquid it receives, the condenser is always empty of liquid and some gas inevitably enters the float chamber. This must be allowed to escape to the evaporator side by a pipe containing a plate pierced by a small orifice.

How is correct regulation ensured when a high pressure float is used?

The charge of refrigerant must be correct; since the condenser is drained, practically all the charge is in the evaporator. Provided that the charge is not excessive, the desired equality between rate of admission to evaporator and rate of evaporation is automatically ensured as the refrigerant evaporated is passed on by the compressor to the condenser, which in turn passes it all to the evaporator via the high pressure float. Too small a refrigerant charge results in part of the evaporator surface being wasted; too great a charge causes liquid to enter the compressor.

What are the limitations of the high pressure float?

(1) If a plant is operated under frequently varying conditions, frequent adjustment of the refrigerant charge is necessary.

(2) Several evaporators in parallel cannot be fed by high pressure float regulators as they cannot distribute liquid correctly.

What is a low pressure float?

It is similar to a high pressure float with the exception that the valve is on the inlet side; hence the float chamber is at evaporator pressure. The liquid level in the chamber, which is held constant by means of a float, may be the same as that in the evaporator coils, or may be higher, providing a head which causes a circulation through the coils several times as great as the rate of evaporation. The low pressure float is thus a level maintaining device for use with flooded evaporators. Several evaporators in parallel can be served by low pressure floats.

What are the disadvantages of the low pressure float regulator?

(1) It necessitates a large refrigerant charge.

(2) Although it permits the presence of a reserve of refrigerant in the liquid receiver, this would be a disadvantage

should the valve stick open, allowing liquid to flood back to the compressor.

What is a capillary tube regulator?

It consists of a length of small bore tubing, the frictional resistance of which provides the necessary pressure drop. It is sometimes used in small automatic refrigerators, having the advantage of simplicity. Providing the refrigerant charge is correct, it operates to some extent automatically; thus if the condenser is completely drained of liquid so that vapour as well as the liquid enters the capillary tube, the pressure drop through the latter is increased and the rate of entry of refrigerant into the evaporator is reduced, so that some of the charge is transferred to the condenser again.

11

PIPING AND PRESSURE DROP

How should pipe diameters be determined?
In most cases the criterion should be the avoidance of excessive pressure drop.

What is the disadvantage of pressure drop?
In delivery pipe lines, it involves an increase in power and a decrease in volumetric efficiency; in evaporators, the mean temperature difference is decreased, thus reducing the heat transferred by a given surface; in suction lines, the compressor suction is lower than that required in the evaporator and the refrigerating duty of the machine is therefore decreased. In the case of the circulation of brine or water, pumping power is increased and a larger size of pump may be necessary.

What factors determine the pressure drop in a given case?
The velocity of liquid (or gas), its density and viscosity, the internal diameter of the pipe and degree of roughness of the internal surface and the number and shape of bends, etc.

What is viscosity?
The viscosity of a liquid is a measure of its freedom of movement. Thus water has a low viscosity whilst a syrup or a heavy oil will have a high viscosity. The viscosity varies with temperature and will, in general, increase with decrease of temperature.

Viscosity can be measured as the time in seconds taken for a measured quantity of liquid to pass through a very small hole. Thus the lower the viscosity the less time a given quantity will take to pass through. The quantity passed and the size of hole are standardised on the Redwood system and viscosities are frequently expressed as so many seconds Redwood. Thus a light fuel oil might be 70 s Redwood whilst a much heavier one, and therefore much more viscous, could be 700 s Redwood, the definition always being followed by the temperature at which the viscosity was measured.

The absolute viscosity is the load which will produce unit velocity difference between two parallel planes set in the liquid at unit distance apart. If units of pounds and feet are used the viscosity is expressed in lb/h ft. There is no special name; in the metric system the units would be in g/s cm. Quite often 1/100 of this value is used as a unit of viscosity and is named a *centipoise* (cP). 1 cP = 2·42 lb/h ft.

On what does the relationship between pressure drop and velocity depend?

If the flow is truly streamlined without any turbulance (a rather rare occurrence) the pressure drop is proportional to $(\text{velocity})^{1\cdot8}$. As a practical method it is quite usual to assume the pressure drop to be proportional to the square of the velocity, this is easier to calculate and has the merit of giving a result on the high side rather than on the low, which is always an advantage with this sort of calculation.

How can pressure drop be determined?

Assume a turbulent flow, then

$$p = fv^2w/d, \text{ where}$$

p = the pressure drop per 100 ft of straight pipe (lbf)
v = the velocity (ft/s)
w = the density (lb/ft^3)
d = the internal diameter of the pipe (in), and
f is a factor between 0·001 and 0·007 and depends on the Reynolds number. If p is the pressure drop per 100 m of straight pipe in kg, then
$p = 26\,fv^2w/d$, where v is in m/s, w is in kg/m^3 and d is in mm.

What is the Reynolds Number?
The Reynolds number (*Re*) is given by the equation: (*Re*) = Velocity × Pipe diameter × Density/Viscosity, where the units of viscosity are lb/h ft. It has the same value for any particular case whatever set of *self-consistent* units is employed, i.e. it is a dimensionless number. A graph of the Reynolds number is shown in Fig. 12.

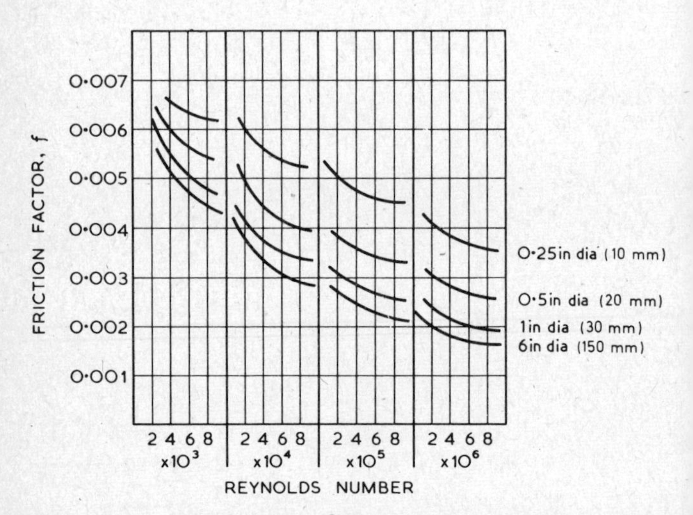

Fig. 12. Graph of Reynolds number and friction factors (f) for steel and wrought iron pipes.

NOTE. The nearest standard metric sizes of pipe are given in parentheses.

How does pressure drop for a given volume of flow depend on pipe diameter?
For streamline flow it is inversely proportional to the fourth power of the internal diameter; for turbulent flow the power involved is between the fourth and the fifth. Therefore reducing a pipe diameter from 2 in to 1 in will, in the case of turbulent flow, increase the pressure drop about 30 times.

In practice, what else must be taken into account in calculating pressure drop in coils, etc.?
The curvature of the pipe, sudden changes of direction or of velocity and the resistance of valves and other obstructions.

How is curvature allowed for?
The actual length of pipe in the curve should be multiplied by a factor to give the equivalent length of straight pipe. This factor depends upon the ratio of the mean diameter of the pipe to the mean diameter of the bend and is approximately 10 times that ratio. Thus if a pipe was of 0·5 in (12·5 mm) mean diameter and was bent to a mean radius of 1·5 in (37·5 mm) the ratio would be 0·5/1·5 or 12·5/37·5 = 0·33, and the multiplying factor would be 3.

How are detachable pipe joints made?
On small pipes of copper, flared connections may be used; otherwise and always where steel pipe is concerned, tongued and grooved flanges welded to the pipe are most satisfactory.

How can stop valves be allowed for?
An open sluice valve offers negligible resistance but an open globe valve of size d in is equal to 25 d ft of pipe. If the diameter is in cm the resistance is equal to 3 d m of pipe.

What materials are used for joint rings?
Usually copper for carbon dioxide, methyl chloride and the fluorocarbons; aluminium or compressed asbestos fibre for ammonia. Copper must not be used for ammonia, or aluminium for methyl chloride. When considering the use of other materials care is necessary to make sure that the material selected cannot react chemically with either the refrigerant or the lubricating oil.

What are the general physical properties of the more commonly met liquids, vapours and gases?
These are listed in Table 4.

Table 4. Some properties of liquids, vapours and gases

	Density (*lb/ft*^{3})				*Specific heat*				*Viscosity* (*lb/h ft*)			
	Liquid		*Vapour and gas*		*Liquid*		*Vapour and gas*		*Liquid*		*Vapour and gas*	
Temp. (°*F*)	−40	+40	−40	+40	−40	+40	−40	+40	−40	+40	−40	+40
Air	—	—	0·09	0·08	—	—	0·24	0·24	—	—	0·036	0·041
Ammonia	43	39	0·04[2]	0·31[2]	1·06	1·11	0·54[2]	0·65[2]	—	0·57	0·024[2]	0·029[2]
Carbon dioxide	69·6	56	1·64[2]	6·93[2]	—	—	—	—	—	0·23	—	0·043[2]
Ethyl alcohol	52·6	50·1	—	—	0·48	0·54	—	—	—	4	—	—
Hydrogen	—	—	0·006[3]	0·005[3]	—	—	3·4[3]	3·4[3]	—	—	0·018[3]	0·020[3]
Dichlorodi-fluoromethane	94·6	90·9	0·256[2]	1·260[2]	0·207	0·227	0·13[2]	0·15[2]	1·02	0·69	0·021[2]	0·029[2]
Methyl chloride	64·4	59·4	0·078[2]	0·437[2]	0·35	0·37	0·18[2]	0·21[2]	—	0·70	0·022[2]	0·028[2]
Glycerol	80·4	78·5	—	—	—	0·54	—	—	—	24[1]	—	—
Ethylene glycol	—	—	—	—	—	0·55	—	—	—	100	—	—
Water	—	62·4	0·059[3]	0·049[3]	—	1	—	0·465	—	3·7	0·018	0·022

NOTES:
(1) Aqueous soln. 50% by wt.
(2) Saturated vapour.
(3) At 1 atm.

	Density *(kg/m³)*				*Specific heat*				*Viscosity* *(cP)*[4]			
	Liquid		*Vapour and gas*		*Liquid*		*Vapour and gas*		*Liquid*		*Vapour and gas*	
Temp. (°C)	*−40*	*+4·4*	*−40*	*+4·4*	*−40*	*+4·4*	*−40*	*+4·4*	*−40*	*+4·4*	*−40*	*+4·4*
Air	—	—	1·49	1·26	—	—	0·24	0·24	—	—	0·015	0·017
Ammonia	689	625	0·64[2]	4·96[2]	1·06	1·11	0·54[2]	0·65[2]	—	0·23	0·010	0·012
Carbon dioxide	1120	897	26·2[2]	112[2]	—	—	—	—	—	0·095	—	0·018
Ethyl alcohol	842	801	—	—	0·48	0·54	—	—	—	1·65	—	—
Hydrogen	—	—	0·103[3]	0·08[3]	—	—	3·4[3]	3·4[3]	—	—	0·007	0·008
Dichlorodi- fluoromethane	1516	1455	4·1[2]	20[2]	0·207	0·227	0·13[2]	0·15[2]	0·42	0·28	0·009	0·012
Methyl chloride	1025	953	1·25[2]	7[2]	0·35	0·37	0·18[2]	0·205[2]	—	0·29	0·009	0·011
Glycerol	1285	1257	—	—	—	0·54	—	—	—	10	—	—
Ethylene glycol	—	—	—	—	—	0·55	—	—	—	41	—	—
Water	—	1000	0·95[3]	0·78[3]	—	1	—	—	—	1·53	0·007	0·009

NOTES:
(2) Saturated vapour.
(3) At 1 atm.
(4) $1\ cP = 1 \times 10^{-2} g\ s^{-1}\ cm^{-1} = 10^{-3}\ Ns\ m^{-2}$, see page 69.

12

COLD STORAGE AND INSULATION

What is the usual temperature for storage of foodstuffs?
Much depends upon the exact state of freshness of the food when put into the store and on the care taken in stowing and housing. The recommendations in Table 5 are mainly those of the International Institute of Refrigeration and do *not* apply to deep frozen produce.

Table 5. Recommended storage temperatures for foodstuffs

Commodity	*Temperature* °F	°C	*Storage period*
Bacon	28	−2	1 month
Beef, chilled	30	−1	2 to 3 weeks
Beef, frozen	14	−10	6 to 8 months
Beer, bottled	40	4	3 to 5 months
Butter, fresh	32	0	4 weeks
Butter, frozen	14	−10	6 months
Eggs, shell	32	0	6 months
Egg powder	50	10	6 months
Fish, frozen	−5	−20	12 months
Hams, green, frozen	18	−8	3 months
Lard	32	0	4 months upwards*
Margarine	32	0	1 to 8 months*
Mutton, chilled	30	−1	10 days
Mutton, frozen	14	−10	6 months
Pork, chilled	30	−1	1 to 2 weeks
Pork, frozen	14	−10	6 months
Pork, salt	40	4	4 to 6 months
Poultry, fresh	32	0	1 to 3 weeks

* Varies according to variety.

What is the practice with regard to humidity in cold storage?
A relative humidity of 80% to 85% is suitable for meat, 85% to 90% for eggs and about 85% for most vegetables. In most cases no attempt is made to provide a means of varying and controlling humidity; satisfactory conditions for short storage are obtained provided that the difference between air temperature and cooler surface temperature is suitable.

How does the difference between air temperature and cooler temperature affect humidity in the chamber?
The smaller the difference between the air and cooler temperatures the higher the relative humidity.

What are the disadvantages of an unduly high and an unduly low relative humidity?
A high humidity favours the growth of moulds on food. A low humidity results in the drying of the food, changed appearance and loss of weight.

What are suitable temperature differences, air to cooler?
For meat 15°F (8°C), for fruit and vegetables 12°F (7°C) and for eggs 8°F (5°C). Forced air circulation is assumed.

What other conditions affect storage results?
Circulation of air. In the great majority of cases a forced circulation of air is essential to maintain a reasonably uniform temperature throughout the chamber. The absence of air movement promotes air stagnation and encourages the formation of mould and slime.

Can fish be stored in a chamber along with other foodstuffs?
Only if it is isolated by being kept in a closed compartment; this may be a galvanised steel tank fitted with a lid, or preferably, with a separate door opening outside the chamber.

What is the usual form for the cooling equipment to take?
Air is cooled by a gilled tube air cooler and circulated around the cold space by means of a fan or fans. Where the space to be cooled is less than 25–30 ft (8–9 m) wide or long,

the fans provided on the air cooler will have sufficient "throw" to send the air all around the space. Where larger areas are concerned the air must be carried by ducts to the furthest parts and discharged there. Intermediate discharge ports are also used where necessary to secure a uniform distribution. Usually the air is allowed to find its own way back to the air cooler suction opening but sometimes special return air ducts are provided.

How should the fan be controlled?

This depends upon circumstances. Usually the fan is started and stopped automatically as the air cooler itself is started and stopped, but there are some cases—the storage of cheese is one—where it is not desirable to permit the establishment of stagnant conditions and the fans must therefore run continuously. A compromise is sometimes made by reducing the fans to half or quarter speed instead of stopping them. This secures sufficient air movement to avoid stagnation but avoids the cost of full speed operation.

What air speed is permissible?

The higher the velocity of the air in a duct the greater will be its resistance and the higher the fan horsepower required. It has to be remembered that the heat equivalent of the horsepower used by the fan is almost entirely passed into the air and it has to be subsequently removed by the refrigerating plant. Sometimes this can be an appreciable amount. On the other hand, the large ducts necessary to secure very low air velocities—and therefore low friction—are themselves costly and take up valuable space. Sometimes, too, actual physical considerations dictate the size of the duct; more often a compromise design is made to achieve a velocity of approximately 1200 ft/min (400 m/min) in the duct.

How can the volume of air passing through an air cooler be determined?

A satisfactory quantity can be found from the formulae:

$$V = D/9, \text{ where}$$

V is the volume (ft^3/min) and

D is the evaporator duty (Btu/h)

and $V = D/8$, where

V is the volume (m^3/min) and

D is the evaporator duty (kcal/h).

An exact calculation involves the use of a psychrometric chart, the heat difference in the air coming in and in the air going out being equated to the evaporator duty allowing a temperature drop through the cooler of 6°–7°F (about 3°–4°C).

Refer also to Section 15 dealing with the cooling of air.

In order to calculate the evaporator surface how should the temperature difference be assessed?

The temperature of the air in the store will be approximately the same as that of the air returning to the cooler. Thus, for a room temperature of 35°F (1·7°C) and evaporating at 20°F (−6·7°C) with a temperature drop over the cooler of 6°F (3°C), the logarithm of the mean temperature difference of 15° and 9°F (8·4° and 5·4°C) is required, i.e. 35° − 20°F and 29° − 20°F (1·7° − (− 6·7)°C and (− 1·3°) − (− 6·7)°C).

On what should the total duty of an air cooler be based?

On the sensible heat duty plus heat leakage through the fabric of the cold room plus, too, the heat given off by electric lamps or other equipment within the cold space. An allowance has also to be made for warm air entering whilst doors are open, usually about 15% of the previous total.

What should be used as the internal lining of a cold chamber?

For small cabinets, vitreous enamel or glazed asbestos cement sheets and sometimes galvanised iron or steel plates. For larger stores, white cement with a concrete floor. The floor should be reinforced in the walkways or where heavy damage can be expected (the dropping of milk churns for instance). The use of a polyurethane insulant bonded directly to plastic-coated steel plates is coming increasingly into favour; in any case it is advisable to ensure proper air circulation by permanent cargo battens on the walls and adequate dunnage on the floors.

How should a chamber be insulated?
Cork slab has been the insulant most used for very many years but it is now giving way to expanded rubber, polyurethane and cellular polystyrene. Lightweight concrete is also used for the higher temperatures, that is to say, for temperatures above freezing point.

What is the thermal conductivity of insulating materials?
Table 6 gives the conductivity of many insulants and of some building materials.

Table 6. Heat conductivity of some materials

Material	*Density*		*Heat conductivity*	
	(lb/ft³)	*(kg/m³)*	*(Btu in ft⁻² h⁻¹ °F⁻¹)*	*(kcal m m⁻² h⁻¹ °C⁻¹)*
Dry brick wall	110	1762	8·0	1·0
Cellular concrete	40	641	1·0	0·125
Cellular concrete	60	961	1·8	0·22
Cork, baked slab	7	112	0·26	0·032
Cork, baked slab	9	144	0·30	0·037
Cork, baked and regranulated	5–6	80–96	0·27	0·033
Expanded rubber	4–6	64–96	0·20	0·025
Glass wool	5	80	0·23	0·028
Mineral wool, felted	4	64	0·23	0·028
Slag wool, felted	8·5	136	0·23	0·028
Insulating wallboard	20	320	0·40	0·050
Polystyrene	1·5	24	0·23*	0·028*
Polyurethane	2	32	0·12*	0·015*

*Can vary with the method of manufacture.

How much insulation should be used?
This is a complex matter. The cost of the purchase and application of the insulant is equated to the saving in running cost as a result of the reduced heat leakage due to the extra insulation. It has to be remembered that the cost of applying the insulation is a high proportion of the total cost and that it costs no more in labour to use, say, a 3 in (75 mm) slab than it does to use a 2 in (50 mm) slab.

Another consideration is that the amount of insulation governs the rate at which the cold chamber will rise in temperature after refrigeration is stopped and this may well be the determining factor since some Insurance Companies insist on some maximum rate of rise each 24 h, usually not more than 3°–4°F (about 2°C).

Yet another factor is that of sweating on the hot side; if the insulation is too thin the outside of the wall will be cold and condensation will take place on it. This can also take place when a cold room, properly insulated for, say, 35°F (2°C), is used for some lower temperature.

What ambient temperature should be assumed in calculating the heat loss?

Two methods may be used; firstly, the mean summer temperature (i.e. the 24 h mean) may be taken, or secondly a higher temperature may be taken to meet the worst conditions ever likely to be met.

In the first case the size of the refrigerating plant should be calculated so as to run for about 15 h each day, assuming automatic control. In the second case the plant would be chosen for a running time of 21 h per day. It has to be remembered that because the product load in refrigerating duty always exceeds the heat loss through the walls, it will be necessary to rate the plant on the sum of the two duties, i.e. the product load plus the loss on whichever method mentioned previously is used.

What is a vapour barrier?

Water vapour will always travel from an area of high vapour pressure to one at a lower vapour pressure; and since vapour pressures are related to temperature it follows that water vapour will try to pass from the hot side of a cold store wall to the inside. This does not matter where the temperatures being maintained are above freezing, but when this is not so the vapour will freeze within the insulation and eventually push it away from the wall. To prevent this an effective vapour seal must be applied to the hot side (usually a bitumen) unless the insulant itself is completely non-hydroscopic, like expanded rubbers or expanded ebonite.

How is the heat loss calculated?
From the relationship

$$I_L = 24AkT_D/d, \text{ where}$$

I_L = Insulation loss per day
A = Surface area
k = Thermal conductivity of the material
T_D = Temperature difference, and
d = Thickness of insulation
the units used being those in which k is expressed.

How is the product load calculated?
The average daily weight of goods housed multiplied by the specific heat and by the temperature range through which it is to be cooled gives the product load. Allowance must be made for cooling containers and for latent heat where the temperature range passes through the freezing point.

If the daily input cannot be ascertained, what quantity may be assumed?
No assumption is safe. The daily input is vital to the sizing of the plant, and the amount to be housed per day must form part of the specification for the store. It must be precisely stated.

What must be added to the insulation loss to determine the total duty?
The heat introduced by opening doors etc., and that to be removed from goods (the product load). If fruit, vegetables or cheese are stored the heat produced by these must be allowed-for. An allowance is also required for electric lights and any motors situated within the store.

What are the specific heats for common foods?
These are summarised in Table 7.

Table 7. Specific heats for common foods

Food	*Specific heat*	*Food*	*Specific heat*
Apples	0·92	Beef	0·75
Butter	0·55*	Lard	0·54*
Fresh cream	0·90	Pork (chilled)	0·51
Poultry	0·80	Veal	0·70

* Varies with the composition of the food and the blending.

The values quoted are those before freezing. After freezing they are considerably less.

If the chamber is held below freezing point and unfrozen foods are put in, what allowance must be made?
The latent heat of freezing must be added to the refrigerating duty required.

How can the latent heat be determined if the specific heat is known?
Approximately, Latent heat = Specific heat × 144 (Btu/lb), or, Latent heat = Specific heat × 79 (kcal/kg). Therefore, to cool meat from 65°F (18·3°C), freeze it (freezing point 30°F (−1°C) approx.) and cool it a further 15°F (8·4°C) which requires the removal of 0·75 (65 − 30) + (0·75 × 144) + 0·4 (30 − 15) = 140 Btu/lb. If metric units are used the required amount of heat to be removed will be 0·75 (18·3 + 1) + (0·75 × 79) + (0·4 × 8·4) = 76·88 kcal/kg. The 0·4 term is the specific heat after freezing which for most foods is about half of that before freezing.

What is frost heave?
Frost heave is the name given to the movement of ground as a result of being frozen because of insufficient insulation underneath a cold store. It is very serious indeed and can cause collapse of the building; once started it is difficult and very costly to stop and is far better avoided altogether.

How can frost heave be avoided?
Where possible the cold store should not be located at or below ground level; even if the lowest floor is no more than

loading bank height above ground the air space thus provided will be sufficient to prevent frost heave provided that suitable precautions are taken where any columns pass into the ground. Where, however, it is necessary for a cold space to be at ground level a really effective frost mat should be interposed between the underside of the floor insulation and the soil or concrete topping.

Frost mats are usually electrically heated and consist of really stout stainless steel heating wires supplied with low voltage current. Sometimes a hot liquid is circulated through a piped grid under the floor.

However the mat is arranged it should be designed to maintain a temperature of around 40°F (5°C) at the actual ground surface.

What precautions should be taken?

Clearly the result of a failure of the mat would be—or could be—serious and also such a failure might well pass unnoticed. There should be an ammeter in the circuit which will at once show whether there is any difference in the current flowing; the mat should be arranged in an interlocking section pattern thus ensuring that the failure of any one section will allow some heat still to be provided by the adjacent section and, finally, the transformer supplying the low voltage should be tapped so as to provide some measure of control. A thermometer buried in the ground will give the soil temperature; the temperature of the mat can be varied as found to be necessary by increasing or decreasing the voltage applied.

13

FREEZING OF FOODS AND ICE MAKING

What is the advantage of freezing foods?
At temperatures well below freezing point, most foods can be stored for very much longer than at higher temperatures.

How far below freezing point must the temperature be lowered?
For long storage, temperatures of 15°F (−10°C) or lower are required, depending on the nature of the food. A temperature of 15°F is low enough for most meat provided that it is not required to store it for longer than about 6 months. At least −5°F (−20°C) is necessary for fish. Consistancy of storage temperature is most important.

Should foods be frozen quickly or slowly?
In nearly all cases, the more rapidly the food is frozen the better its condition on thawing will be.

What are the advantages of quick freezing?
When freezing is rapid, the ice crystals formed are minute; with longer freezing these crystals grow to larger dimensions and thereby cause mechanical damage to the tissues of which organic matter, whether animal or vegetable, is composed. Slow freezing also results in physico-chemical changes, and as a result, water separates from protein and is not completely re-absorbed subsequently.

What must be the time of freezing for a food to qualify as quick frozen?
No precise line of demarcation can be drawn. It is generally considered that a "quick frozen" product has passed

through the temperature zone in which the ice crystals are formed in less than $1\frac{1}{2}$ h. The actual time taken is often very much less than this particularly where the quick freezing process forms part of a production line. The size of the product has a bearing, too, since clearly a pea can be frozen much faster than, say, a piece of meat such as a "steaklet".

How may quick freezing be effected?
There are many methods, which may be divided into three main classes. These are freezing by rapid movement of air (blast freezing), by immersion in a liquid medium at low temperature and by spraying with cold liquid or by contact with refrigerated metal surfaces with or without mechanical pressure.

Various combinations of these methods are often used. Recently there has been considerable development in quick freezing using liquid nitrogen; these units often form part of a production line with arrangements for the recovery and re-liquefication of the nitrogen following its vaporisation.

At what temperature should quick frozen fruit and vegetables be stored?
Preferably below −5°F (−20°C).

What is freezer burn?
Surface damage due to excessive drying during freezing. It has no connection with the burn experienced when the unprotected flesh comes into contact with very cold surfaces. This "cold burn" is similar to a burn received by coming into contact with very hot surfaces, and is caused by the same effect—the rapid passage of heat either into (hot burn) or out of (cold burn) the flesh. For some reason not clearly known a cold burn is much more difficult to heal than is a hot burn.

What is the Zarotschenzeff freezing process?
This utilises a fog or spray of brine at low temperature.

What is "dry ice"?
Dry ice is frozen carbon dioxide, sold under various trade names, such as *Cardice* and *Dricold.* It has the property of passing directly from the solid to the gaseous state without becoming a liquid (sublimation).

How is ice obtained?
The ice cube unit is a frequent accessory of household and other small refrigerated cabinets. It consists of a metal casing either having refrigerant piping sweated to it or made hollow with refrigerant passages, and provided with shelves on which the ice trays, usually made of aluminium, rest. Small special cabinets for making ice cubes only are also available in many sizes; they are used especially in hotels, bars, clubs, etc.

Where very large amounts of ice are required the water is frozen in galvanised cans immersed in a tank of cold brine. Automatic ice makers are also available which provide a continuous supply of ice in flake or ribbon form. They usually have a capacity of 2 tons (2000 kg) of ice per day and upwards.

How much refrigeration is required to produce one ton of ice?
The refrigerating duty is made up of the heat to be removed from the water in cooling it to freezing point, the latent heat of freezing (144 Btu/lb, 79 kcal/kg) the heat to be removed in cooling the ice to approximately brine temperature, insulation loss from brine tank, and propeller power. The total varies with the size of tank, amount of insulation, brine temperature and initial water temperature.

What must be done if clear ice is required?
Agitation of the water in the cans by injection of air must be provided. There are two methods, firstly the high pressure system in which air at between 15–30 lbf/in^2 (1×10^5 – 2×10^5 N/m^2) is introduced at the bottom of the cans, and a low pressure system using air at about 2 lbf/in^2 ($1{\cdot}4 \times 10^4$ N/m^2) introduced into the top of the cans by a small tube which is removed before the contents are quite frozen.

14

DAIRY REFRIGERATION AND ICE CREAM

What is the advantage gained by cooling milk and to what temperature must it be cooled?
The multiplication of harmful bacteria, which proceeds at a phenomenal rate at medium and high temperatures, is greatly slowed down at temperatures around 40°F (4°C), and this is the usual temperature to which milk is cooled and at which it is stored.

How is milk cooled?
The usual method is by passing it over a surface cooler consisting of horizontal tubes of tinned steel, copper, or stainless steel. The cooler usually consists of two sections, the upper one having water passing through it and the lower one either brine or evaporating refrigerant. Coolers are made in sizes from 25 gal/h (113 l/h) upwards.

How much water is required?
It is good practice to use 3 gal (13·5 l) of water for every gallon of milk. Water must flow upwards through the cooler.

How much heat has to be removed from the milk in cooling it through a given temperature range?
About 9·6 Btu gal^{-1} $°F^{-1}$* or about 1 kcal l^{-1} $°C^{-1}$.

To what temperature is the milk cooled on the water section?
Sufficient surface should be provided to cool the milk to within 5° or 6°F (3°C) of the water inlet temperature.

* The specific heat of cream varies considerably with the temperature but an average figure of 9·6 Btu or 1 kcal per gallon or litre respectively is approximately correct for the temperature range 140°–40°F (60°–4°C).

What evaporating temperature is required for the refrigerating section?
About 28°F (−2°C).

What are the advantages and disadvantages of direct expansion and brine milk coolers?
With a direct expansion cooler the maximum duty is obtained from the refrigerating machine, as the evaporating temperature is higher than when brine has to be cooled as an intermediary. Also, the installation is simpler and the accessories less costly.

With brine, a smaller machine can be used and made to cool a tank of brine when milk is not being cooled, in order to provide a reserve of refrigeration which is called upon during milk cooling.

Can a milk cooler and a milk storage chamber be satisfactorily dealt with by one machine?
Yes. If the cooler is of the water and direct expansion type, the machine will run alternately on this and on the evaporator in the cold room. The rise of temperature of the cold room during the milk cooling period will not normally be serious. If a brine cooled cooler is used, the tank room may be placed behind a casing with the fan in the cold room, and it will serve to cool the latter. In addition an external insulated tank may be used and brine circulated to both the milk cooler and the cold room.

What capacity of store is necessary to hold a given amount of milk?
About 1·5 gal/ft^3 (24 l/m^3) gross if bottled and crated.

In calculating the refrigerating duty for a pre-cooled milk store what cooling load must be taken into account?
The heat to be removed from bottles and crates. If the milk has already been cooled to 40°F (4°C) and is to be stored at the same temperature, good practice would be to allow for cooling the milk through 2°–3°F (1°–2°C) in the chamber and for cooling bottles and crates from 70°F (21°C). Basing these figures on the use of pint bottles and metal crates of

average weight, the duty involved in cooling the bottles and crates amounts to approximately 75 Btu (19 kcal) for each gallon of milk stored.

What fitting is it advisable to instal in the refrigerant circuit when a direct expansion milk cooler is used?

A back pressure valve should be fitted in the suction line. This automatic valve, rather like a constant pressure regulator, prevents the pressure on its *inlet* side from falling below the level at which it is set, by throttling whenever there is a tendency for this to happen. It is used in conjunction with an ordinary thermostatic regulator and for a milk cooler should be set for a minimum cooler pressure corresponding to 28°–30°F (−1°– −2°C) thus preventing the icing of the cooler which otherwise would readily occur as a result of the rate of milk flow over the cooler being reduced at times.

If water is scarce but reasonably cool, how may economy in its use be effected?

The water may be passed through the condenser of the refrigerating machine after leaving the milk cooler—that is assuming that only a water and direct expansion milk cooler has to be dealt with.

What is ice cream mix?

The mixture of ingredients which after freezing and aerating becomes ice cream. It usually consists of varying proportions of milk, fat, butter fat, other milk solids, sugar and a small quantity of stabiliser—frequently gelatine.

What is homogenising?

After the mix is prepared and pasteurised (held at 145°F (63°C) for 30 min) it may be treated in a homogeniser, which subjects it to very high pressure and passes it through narrow passages at high velocity, thus breaking up the fat globules into smaller ones, leading to a smoother product.

What is the next stage?

The mix is cooled quickly, generally on a surface cooler, to 40°–45°F (5°–7°C).

How does a mix cooler compare with a milk cooler?
It must be provided with more surface because of the higher viscosity and the consequent lower rate of heat transmission of the mix; how much more depends greatly on the composition of the mix.

What happens to the ice cream mix in the freezer?
It is partially frozen and at the same time air is whipped into it.

Into what classes may ice cream freezers be divided?
Two, batch freezers and continuous freezers.

How is a small batch freezer made?
In a simple form for small outputs, it consists of a brine tank in which is situated a revolving ice cream can with fixed scrapers and revolving central beater. The can is surrounded by evaporator coils which cool the brine.

How is a larger batch freezer made?
With a fixed cylinder, horizontal or vertical, in which revolving scrapers are fitted. Brine is circulated through a jacket, or the freezer may be of the direct expansion type, with primary refrigerant in the jacket. It is usual to stop the refrigeration before whipping is completed.

What brine or evaporating temperature is required?
Usually −5°– −10°F (−21°– −24°C).

If the freezer is of the direct expansion type, using methyl chloride or a fluorocarbon, what features must be present in its construction?
Either passages must be arranged in the jacket to ensure a continuous flow of refrigerant in order to carry oil back to the compressor, or, if the freezer is of the flooded type, an effective method of distilling the refrigerant off from the jacket and returning the oil to the compressor crankcase must be adopted.

At what temperature does ice cream mix freeze and at what temperature must it be removed from the freezer?
It freezes over a considerable range of temperature, starting

at about 27·5°F (−2·5°C). The temperature of the ice cream when withdrawn from the freezer is about 23°F (−5°C) in the case of a rotating can batch freezer, a little higher in a modern type of batch freezer, whilst a continuous freezer, by virtue of the air pressure used, makes it possible to produce a stiffer ice cream at a temperature of around 21°F (−6°C).

What are the special characteristics of a continuous freezer?
Ice cream mix and air in controlled proportions are pumped continuously through the freezing cylinder under considerable pressure.

Is the heat removed from the ice cream the total refrigerating duty required?
No. The heat leakage into the freezer and the power absorbed by the mixing device must also be taken into account.

What is over-run and what is the relationship between weight and volume of ice cream mix and ice cream?
Over-run, or swell, is the increase in volume of the mix during freezing as a result of the incorporation of air. It usually varies from about 60% to 90%, the higher figures being obtainable with a modern continuous freezer. One gallon of mix weighs about 11 lb (5 kg), so that 1 gal of ice cream weighs, say, from 11/1·6–11/1·9 lb (5/1·6–5/1·9 kg), i.e. from 6·9–5·8 lb (3·12–2·62 kg) according to the extent of the over-run.

How is a bulk hardening and storage room cooled?
The cans are usually stood on shelves made from evaporator grids and carrying metal plates, further grids are fitted to walls and ceiling.

How is a brick hardening and storage room cooled?
Usually in exactly the same way as a bulk room; in this case, however, a vigorous air circulation considerably increases the rate of hardening and the installation of a blast freezer, similar to those used for quick-freezing foods, would be advantageous where a large output has to be hardened as quickly as possible.

How is an ice cream conservator made and what is it used for?
Generally, it consists of an insulated metal tank with direct expansion piping in the form of grids sweated to the outside and covered with insulation. The heat leaking into the insulation from outside is therefore absorbed by the cold sides of the tank without ever reaching the interior. This type of cooling is excellent where very little heat is produced inside the storage space, or introduced into it through opening the lids and is therefore well suited to the storage of ice cream already hardened.

What temperature is maintained in the conservator?
Usually around 15°F for bulk ice cream and 0°F for bricks (−10°C and −18°C).

What method of hardening ice cream in small quantities can be used apart from a hardening room?
A conservator is possible but should be made with a brine jacket in which the evaporator coils are immersed. This will help to prevent fluctuations of temperature when ice cream from the freezer is put into the conservator.

15

THE COOLING OF AIR

How does air behave when cooled?
It may be regarded as a mixture of various proportions of dry air (itself a mixture but behaving as a single gas) and water vapour.

What governs the amount of water vapour that can be contained in a given amount of air?
The answer to this question may be made clearer if the question is put in another form, i.e. "what governs the weight of water vapour that can be contained in a given volume of space?" Water, like any other liquid, is capable of exerting a vapour pressure, the amount of which depends solely on the temperature at the liquid surface. If the liquid is in contact with its vapour, the latter is saturated; its pressure therefore corresponds to the temperature and it has a corresponding density or weight per unit volume. Thus, the maximum weight of water vapour which can exist in 1 ft^3 of air depends solely on the temperature; this weight is simply the density of saturated water vapour at this temperature and is quite unaffected by the presence of air in the same space. For example, at 50°F (10°C) the saturation vapour pressure of water is 0·178 lbf/in^2 ($1{\cdot}227 \times 10^3 N/m^2$) and its density 4·12 gr/ft^3 (0·95 g/m^3); while at 80°F (27°C) the pressure is 0·507 lbf/in^2 ($3{\cdot}495 \times 10^3 N/m^2$) and the density 11·1 gr/ft^3 (2·5 g/m^3). These weights are the maximum weights of water vapour which can exist in 1 ft^3 of space at the respective temperatures, whether the total pressure existing is normal atmospheric or any other. It will be seen from

this example that as the temperature rises so does the maximum water content; it is usual to say that air can hold more water vapour at higher temperatures, but really it is space that can do so.

When air has the maximum water content possible at its particular temperature, what name is given to it?
It is said to be *saturated* air, or to have a relative humidity of 100%.

What is relative humidity?
The ratio of the pressure of the water vapour present to the maximum possible water pressure at the prevailing temperature. Thus, if air at 50°F (10°C) contains water vapour exerting a pressure of 0·089 lbf/in^2 ($6{\cdot}137 \times 10^2$ N/m^2) it has a relative humidity (r.h.) of 50% (see previous answer). At moderate temperatures (up to 100°F (38°C) or thereabouts) this ratio is very nearly the same as the ratio of the *weight* of vapour per ft^3 or per m^3 to the maximum possible weight. However, it deviates appreciably from the latter ratio (which is often called *percentage humidity*) at higher temperatures.

What is dry bulb temperature?
Simply the actual temperature of the air as measured by an ordinary thermometer.

What is absolute humidity?
The weight of the water vapour which is associated with a unit quantity of air. Hygrometric tables and charts are, today, usually made on a basis of 1 lb or 1 kg of *dry* air plus its associated water vapour as the unit; and on this basis absolute humidity is the weight of water vapour associated with 1 lb or 1 kg of dry air.

What is the temperature of adiabatic saturation of a given sample of air?
Suppose that a quantity of the air at a given temperature is enclosed in an insulated vessel together with more than sufficient water to saturate it, and suppose that sufficient time is allowed for equilibrium to be reached. The air will

become saturated by evaporation of some of the water; the heat to effect this is obtained from the air which is therefore cooled. Suppose, moreover, that the initial temperature of the water was the same as the equilibrium temperature finally attained of air and water. What has happened, then, is a transfer of heat from air to water, resulting in a reduction in the temperature of the air and the evaporation of an equivalent weight of water. The equilibrium temperature is known as the *temperature of adiabatic saturation* of the air. For a given initial temperature, it is evidently a measure of the initial humidity of the air; the lower this is the more water is evaporated before saturation is reached, and the greater is the reduction in the temperature of the air.

What is the wet bulb temperature of a given sample of air and how does it differ from the temperature of adiabatic saturation?

The wet bulb temperature is that read by a wet bulb thermometer; this is an ordinary thermometer the bulb of which is wetted by being surrounded by a sheath of muslin kept wet by pure water. For reliable readings to be obtained the air must flow past the bulb at a speed of not less than 10 ft/s (3 m/s). Unlike the adiabatic saturation temperature the wet bulb reading is a *dynamic* equilibrium temperature. When it is reached water is evaporating from the wet bulb into the air at a rate depending on the vapour pressure difference (that of the water on the bulb less that of the water in the air) and on a coefficient of diffusion analogous to the coefficient of heat transfer; at the same time, heat is being transferred from the air to the water film at a rate which corresponds to the rate of evaporation and also to the temperature difference (air-bulb) and the coefficient of heat transfer. Thus the wet bulb temperature depends on coefficients of vapour diffusion and of heat transmission (the fact that it is constant for air speeds exceeding 10 ft/s (3 m/s) shows that above this speed both coefficients increase proportionately) while adiabatic saturation temperatures are independent of these coefficients. By very good fortune, the values of the coefficients happen to be such that in conditions usually met with the two temperatures are almost the same and they are often treated as identical. It is a matter of coincidence, and the corresponding temperatures for a liquid other than water or a gas other than air would be by no means equal; nor are they so for air at high temperatures and with high moisture contents.

What is the dew-point temperature of a given sample of air?
The quantity of moisture contained in air having a relative humidity of less than 100% is less than the maximum possible at the existing temperature, but is the maximum possible at some lower temperature. This lower temperature is the dew point, and if the air is cooled to a temperature slightly below the dew point moisture will begin to condense. The dew point of a given sample of air depends only on its absolute humidity; for example, air at atmospheric pressure, 70% r.h. and a temperature of 70°F (21°C) contains 77·5 gr of water vapour per pound of air (11·1 g/kg). This fixes its dew point at 60°F (15·5°C) because saturated air at this temperature also contains this weight of water vapour.

What properties are necessary in order to specify completely the condition of air at normal pressure?
Any two of these: dry bulb temperature, wet bulb temperature, dew point, relative or percentage humidity, or the same with absolute humidity substituted for dew point.

Why should all air treatment calculations be worked out for unit weight of dry air?
Because the weight of dry air is the only quantity which always remains constant when air is subjected to cooling and heating processes. If either unit volume or unit weight of moist air is treated as the unit it will be necessary to make adjustments as the work proceeds for change of volume with temperature, or change of weight with evaporation or condensation.

What is the psychrometric chart and what are its advantages?
It is a graph, the co-ordinates of which are usually either

dry bulb temperature and absolute humidity, or enthalpy (total heat) and absolute humidity. Families of lines are drawn on the graph showing constant dry bulb temperatures, wet bulb temperatures, enthalpies, absolute humidities, relative or percentage humidities, and specific volumes. The psychrometric chart, giving all the necessary properties of air on a single sheet, greatly simplifies air conditioning calculations and makes it possible to plot lines showing at a glance the changes in the condition of air as it passes through various processes.

What are the specific volumes shown on the psychrometric chart?

Not quite the true specific volumes of the air-water vapour mixture but the volume of unit weight of dry air plus its associated water vapour.

What happens when moist air is passed over a cold surface?

This process, simple as it sounds, is in reality the most complex and quantitatively the most difficult to understand of those with which the refrigerating engineer is usually concerned. The temperature of the air is reduced and the rate at which heat is removed from it to effect this reduction depends upon the amount of surface, the temperature difference between the air and the surface, and the coefficient of heat transfer, the latter again varying with air speed. Thus the removal of *sensible* heat is a comparatively straightforward process. Side by side with this, however, there is another process. This is the physical movement of water vapour through the air to the cold surface, which proceeds at a rate depending on the vapour pressure difference (pressure of water vapour in the air less that at the surface) and on the coefficient of diffusion of water vapour through air. Besides the transfer of sensible heat previously described, it is evident that a further quantity of heat must be transferred in order to condense the vapour. This transfer, however, does not meet with appreciable resistance because it takes place at the water surface, not through the stagnant film of air over the water, as in the case of sensible heat.

How does the designer calculate the amount of surface required for given cooling and dehumidification loads?

Complications arise from the fact that the temperature difference and the vapour pressure difference are changing all the time as the air passes through the cooler. The former change may be allowed for by using the logarithm of the mean temperature difference in conjunction with the air cooling (sensible heat) load, thus:

$$\text{Surface} = \frac{\text{Sensible duty per hour}}{\text{Log(mean temp. dif.)} \times U}$$

where U is the coefficient of heat transfer in whatever units are being used, while the latent heat, as explained above, need not enter into the calculation. The problem, therefore, resolves itself into this: having decided the amount of surface and the temperature at which it is to be held (which is practically equal to the evaporating temperature in the case of a pipe cooler) how much dehumidification will be effected? Here again coincidence helps us. It can be imagined how convenient it would be if successive points in the cooling process could be represented by a straight line on the psychrometric chart, i.e. if the ratio of sensible heat removed to moisture condensed remains the same throughout the cooling process.

It can easily be shown that this is the case if two conditions are fulfilled: (*a*) that the fall in vapour pressure is proportional to the fall in absolute humidity (which means that the mathematical law of vapour transfer is strictly analogous to the law of heat transfer), and (*b*) that a particular numerical ratio exists between the coefficients of heat transfer and of vapour diffusion. Now the remarkable thing is that both these conditions are approximately fulfilled over a range of temperatures commonly met with. Thus a straight line drawn on the psychrometric chart closely represents the successive stages of temperature and humidity that the air passes through as it travels over a cooling surface, the temperature of which is uniform.

At what angle must the cooling line be drawn on the psychrometric chart?

This is easily answered by the consideration that if sufficient cooling surface were available, the temperature of the air would eventually be reduced almost to that of the cooler, and its vapour pressure almost to that of the film of water on the cooler. Hence the line must be drawn so as to pass through the point on the saturation curve which corresponds to the temperature of the cooling surface.

When a gilled cooler is used, how does this affect the dehumidification?

In two ways: (*a*) the temperature of the cooling surface is considerably higher than that of the refrigerant; and (*b*) the temperature of the cooling surface varies through the cooler, being highest where the air is warmest.

How may an approximate idea be obtained of the cooling line with a gilled cooler?

By dividing the temperature range into a number of parts, calculating the surface temperature for each part and drawing a series of straight lines to the corresponding points on the saturation line. These straight lines will enclose a curve which approximately represents the cooling process.

How may the surface temperature be found?

By equating the sensible heat transferred to the surface plus the latent heat load to the heat absorbed by the refrigerant, we obtain the relation:

$$\frac{T_1}{T_2}=\frac{H_1 Rr}{H_2}, \text{ where}$$

T_1 is the temperature difference between the surface and the refrigerant

T_2 is the temperature difference between the surface and the air

H_1 is the heat transfer coefficient, air to surface

H_2 is the heat transfer coefficient, surface to refrigerant

r is the ratio of external/internal surface

R is the ratio of total/sensible heat load.

How may air be cooled and brought to a definitely controlled humidity?
By passing it through a washer consisting of water sprays, the water temperature being controlled thermostatically. If several rows of sprays are used, the air leaving the washer will have been cooled almost to the water temperature and almost saturated. Its moisture content will therefore depend only on the temperature of the washer water. If the reduction of the moisture content to the required amount results in too low a temperature, re-heating must be resorted to.

How is the temperature of the washer water controlled?
The best way of obtaining sufficiently close control is by means of a thermostatic modulating valve which automatically mixes the refrigerated water with the necessary proportion of warmer water drawn from the cooler outlet to produce the required temperature.

How can air be delivered to a conditioned space at a relative humidity lower than that obtainable with a direct expansion cooler?
By cooling the air to a lower temperature, such that the dew point of the air leaving the cooler is equal to that required, and then re-heating.

What temperature and relative humidity are desirable for comfort in a conditioned room in summer?
Considerable difference of opinion exists on this point. It is important to avoid too great a temperature difference between the outside and inside air. Suggested rules for application in the U.K. are that the temperature should not be more than 10°F (5·5°C) below that outside and not lower than 70°F (21°C), and the relative humidity between 30% and 70%.

How is the refrigerated load on a conditioned space made up?
Heat from outside, including direct sun radiation (use external blinds where possible); heat from occupants, lighting, machinery etc., and the sensible and latent heats from the cooled air.

How may sun heat be allowed for?
By estimating an average of 180 Btu h^{-1} ft^{-2} (490 kcal h^{-1} m^{-2}) of unshaded window. This figure will not necessarily apply outside Latitude 20° N or S and refers to windows facing E or W.

How much heat is generated by occupants?

Persons seated at rest	400 Btu/h or 100 kcal/h
Engaged on very light work	600 Btu/h or 150 kcal/h
Exertion equal to moderate walk	1000 Btu/h or 250 kcal/h
Exertion equal to fast* walk	1400 Btu/h or 350 kcal/h

* about 4 mile/h (6·4 km/h).

How much fresh air is required?
Approximately 1000 ft^3/h (300 m^3/h) per person. Three times this if smoking is allowed. This figure is the amount of fresh air required and not the quantity actually circulated, which is usually much more than this, the difference being recirculated. The fresh air intake has to be cooled to room conditions.

The whole question of the amount of air is complex and depends very much on conditions because adequate ventilation requires not only a proper amount of fresh air, but the removal of contaminants, heat, smoke, dust, etc., and the maintainance of a level of humidity which will not cause discomfort.

16

LOW TEMPERATURES

What special problems have to be solved in obtaining low temperatures such as evaporating to lower than −40°F (−40°C)?
They are many. The following are among the principal ones: high ratio of compression involving large clearance losses and piston slip losses, very low absolute suction pressures requiring the use of especially light suction valves to avoid inertia losses, high delivery temperatures due to high compression ratios, loss of duty owing to the large temperature range through which liquid has to cool itself in passing through the expansion valve, high viscosity and sometimes breakdown of lubricating oil in cold parts of the circuit and difficulty in operation of expansion valves due to small weight of refrigerant circulating in a given time.

What is compound compression and why is it used?
In *compound* or *stage* compression, the refrigerant is compressed through part of the pressure range in one compressor (or in one stage of a multistage compressor) and then passed to a second compressor, or stage, of smaller swept volume, which carries the compression further (see Fig. 13). Usually two stages only are used but sometimes three. The advantages are: (*a*) the compression ratios of the low pressure cylinders (which form the refrigerating compressor proper) are considerably reduced and the volumetric efficiency, which might in single stage compression approach zero, is thereby made reasonably high; (*b*) the division of compression into more than one stage makes it possible to cool the compressed gas between the stages and so avoid excessive

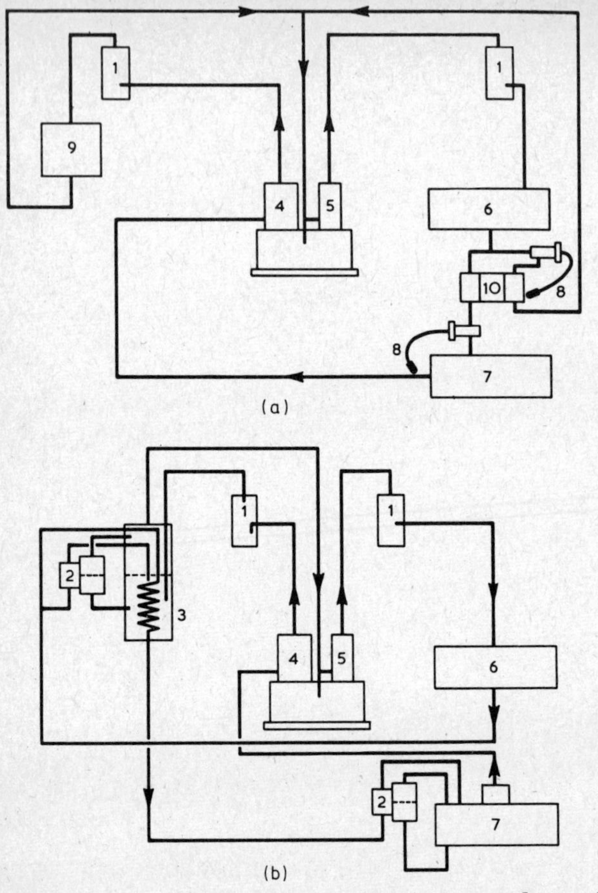

Fig. 13. A typical two-stage refrigerating plant, (a) using a fluorocarbon refrigerant and (b) using ammonia.

1. Oil separator 2. Low pressure float 3. Intercooler 4. Low pressure cylinder 5. High pressure cylinder 6. Condenser 7. Evaporator 8. Thermostatic regulator 9. Gas cooling radiator 10. Liquid cooler

final delivery temperatures; (*c*) by virtue of the interstage cooling, the power consumption is somewhat reduced, and (*d*) at the cost of complicating the refrigerant circuit, matters may be arranged so that the bulk of the liquid is cooled after leaving the condenser by the evaporation of a small part of it at *intermediate* pressure, the vapour so produced being pumped by the high pressure cylinders, which are thereby made to increase the refrigerating duty in addition to their normal task of boosting the pressure of the low pressure delivery up to condenser level.

In what circumstances should gas be cooled between stages?
If the Mollier chart shows that the final delivery temperature would otherwise exceed 300°F (149°C). This is often the case with ammonia, rarely with the fluorocarbons.

How may gas intercooling be effected?
If the low pressure delivery temperature exceeds the temperature of the water supply sufficient cooling may be obtained by the use of water in a shell and coil or multitubular cooler. Otherwise it may be necessary to utilise the evaporation of some of the liquid at intermediate pressure. Sometimes, particularly where the fluorocarbons are concerned, an air-cooled radiator can be used.

What is a cascade system and why is it used?
In this system there are two machines. The low pressure machine works as a refrigerator in the normal way, and the high pressure machine is used to cool the condenser of the low pressure machine, thus enabling the latter to work with a suitably low compression ratio. The two refrigerating systems have quite independent circuits and may even use different refrigerants if desired. The only link between them is the heat exchange unit, which is at the same time the low pressure condenser and the high pressure evaporator.

What are the comparative merits of compound and cascade systems?
The compound system, using one two-stage compressor, evidently requires less machinery; it also takes appreciably

less power. On the other hand cascade working may make it possible to use standard compressors instead of having to design a special one; it simplifies refrigerant circuits and oil problems and permits the use, in the low temperature circuit only, of a refrigerant such as ethane or ethylene, which if condensed at normal temperatures, as in a compound machine, would involve pressures too high for ordinary plant.

What other method is available?
Increasing use is being made of machines constructed on the *Stirling* principle or cold gas refrigeration, especially for the liquefaction of gases. Cold gas refrigeration has been explained on page 30.

What refrigerants are particularly suitable for low temperature work?
Ammonia is quite satisfactory for moderately low temperatures, but below −70°F (−57°C) evaporation or thereabouts, the large volumes necessary and the high temperatures of compression are disadvantageous. Methyl chloride has a very low triple point but its specific volume increases so rapidly with decrease of temperature that it cannot really be considered a low temperature refrigerant except, perhaps, for very small duties. R.12 is quite good for low evaporation temperatures and has low delivery temperatures so that gas intercooling is rarely necessary.

Liquid cooling with the aid of the second stage is however essential, about 5% increase over the duty with liquid at 80°F (27°C) being obtained for each 10°F (5·5°C) by which the liquid temperature is reduced. Although the ideal duty per unit volume of R.12 is less than that obtainable from ammonia, the difference is smaller the lower the temperature, and may be very slight in practice by virtue of the higher volumetric efficiency obtainable with R.12. R.22 is now being used to a considerable extent and gives fairly high refrigerating duty with reasonable delivery temperatures. For very low temperatures (below −110°F (−79°C)) the refrigerants at present used in the vapour compression system are ethane, ethylene and some members of the fluorocarbon group, particularly R.13 and R.14.

What modifications are necessary to a thermostatic regulator to be used for low temperature work?

The bulb should be charged with a fluid having considerably higher saturation pressures than the refrigerant, assuming the latter to be ammonia, methyl chloride or one of the fluorocarbons. Otherwise, the very small pressure change in the sensitive element corresponding to a change of a few degrees in temperature would provide insufficient force to overcome friction in the instrument and very sluggish response would result.

17

AUTOMATIC CONTROLS

What is a thermostat?
It is an automatic switch, the opening and closing of which is actuated by change of temperature. When it is used to control a heater, a fall in temperature makes the circuit. When it controls the motor of a refrigerating plant it must be constructed so as to close the circuit when the temperature rises. A thermostat will pass sufficient current to supply small motors but for larger motors or where the load is inductive, it must be arranged to supply current to a relay switch which in turn when closed connects the motor or apparatus to the mains. Most thermostats used for refrigerating work operate by means of a bellows containing a suitable liquid, or by means of a phial connected to a bellows. The force required to operate the switch is obtained from the changing pressure of the saturated vapour. The switch itself may take the form of a tube containing mercury which makes and breaks contact as the tube tilts, or it may have snap metallic contacts.

What is the temperature range of a thermostat?
The part of the temperature scale over which it will operate. The pressure of the vapour in the phial is balanced by a spring and the amount of adjustment possible in the force exerted by the spring governs the range of operating temperature.

What is the temperature differential of a thermostat?
The difference between the temperatures at which it makes

and breaks respectively. For an ordinary commercial thermostat used to control a refrigerator a differential of about 4°F (2°C) is usual; this can also be expressed as ±°F or ±°C as the case may be. Thus if an average temperature of say, 36°F is required in a cold chamber the machine will start when the temperature reaches 38°F and stop when it reaches 34°F.

Is closer temperature control possible and desirable?

In ordinary commercial work it is possible to reduce the differential slightly, i.e. to $\pm 1\frac{1}{2}$°F or even to ±1°F as a minimum, and an adjustment is provided for this purpose, but little would be gained by so doing. No improvement in the condition of the food stored would be observable, and the frequent starting and stopping of the machine would tend to be detrimental to the compressor, the motor and the starting equipment, particularly where starting resistances are employed.

How is close temperature control obtained if this should be necessary, as, for instance, in scientific work?

Thermostats of special type providing a differential as low as 1/100 of a °F are obtainable but it is not sufficient to control the refrigerating machine by such a thermostat because the mere opening and closing of a circuit does not switch refrigeration on and off like an electric lamp. After the machine stops, in particular, refrigeration continues to be produced as a result of the heat absorbing capacity of the cold evaporator and the refrigerant it contains. The only way to ensure closer control than that afforded by the simple use of a commercial thermostat is to provide a continuous supply of refrigeration (either by direct expansion or by brine) somewhat greater than is required and to balance the surplus by an electric heater or heaters controlled by a thermostat having the required differential. Heaters made of fine wire having an extremely small heat capacity are obtainable, providing a source of heat which can be made available and cut off almost instantly.

How is relative humidity controlled?

Usually by a hygrostat (humidistat) which is a piece of apparatus similar in operation to a thermostat, but actuated by the change in length of some hygroscopic material (often human hair) with changing relative humidity. When dehumidification is required, the hygrostat is arranged to switch on a refrigerating machine or to operate a mixing valve which adjusts the temperature of a secondary refrigerant supplied to the cooler.

What is a dew point thermostat?

Simply an ordinary thermostat, but used in such a way as to control humidity. When air is passed through washer sprays using refrigerated water, its moisture content on leaving depends on the temperature of the water. Consequently a thermostat used to control the temperature of the washer water through the operation of a mixing valve thereby determines the absolute humidity—and so the dew point—of the air leaving the washer.

What is a pressostat?

An automatic switch connected by a small bore tube to the suction of the compressor, stopping the latter when the pressure falls to a certain value and starting it again after a definite rise in pressure. Pressostats may be used in several ways:

(1) Instead of a thermostat, with the object of automatically defrosting the evaporator while at the same time providing some degree of control over temperature.

(2) To prevent the compressor from pumping down to a very low evaporator pressure in the event of a reduction or a cessation of the heat load.

(3) In conjunction with a thermostat, with the object of at least partially transferring the refrigerant charge from the evaporator to the condenser before the machine stops, so as to avoid the return of liquid to the compressor when the latter starts again.

In case (3) the thermostat controls a solenoid-operated stop valve in the refrigerant liquid line and the pressostat controls the machine; thus when the temperature in the chamber has fallen to the required minimum, the thermostat closes the liquid valve and the compressor continues to run for a little while until the evaporator pressure has fallen to a fairly low level. Note, however, that this does not mean that the evaporator is empty or even nearly so.

What is a high pressure cut-out?

A pressure operated switch which stops the machine on the rise of pressure to a level approaching danger, and usually has to be reset by hand. It should be a standard fitting on all automatic machines with water cooled condensers, as a precaution against failure of water supply and also on machines with air-cooled condensers when the condenser fan is driven by an independent motor. It should be set to operate at a pressure of approximately 20 lbf/in^2 ($1{\cdot}38 \times 10^5$ N/m^2) above the highest likely to be met with in normal operation.

What is loss of control in a thermostat?

The pressure in the sensitive element of a thermostat is that corresponding to the temperature of the liquid at its surface, so for proper control the liquid must be at the same temperature as the air, brine or other medium to be controlled. In cold weather, however, the bellows of a thermostat of the bulb and bellows type, if installed in a exposed position, may be cooler than the bulb in the chamber; when this happens, the vapour, which is already saturated, begins to condense in the bellows, being replaced by evaporation of liquid from the bulb, and this may go on until all the charge is transferred to the bellows. The temperature of the bulb will then have no further influence on the operation of the instrument and the machine will not run.

Where should a cold chamber thermostat be fitted?

Preferably near the air returning to the cooler. A thermometer should be installed close to it to check its operation. A thermostat should not be fitted close to a door or in the stream of air leaving a cooler.

How is the flow of water to a condenser automatically controlled?

(1) By a solenoid operated valve, wired so as to open when the current is switched on to the compressor motor and to close when the latter stops. In this case there is a constant rate of flow, controlled by a throttle valve which is set by hand.

(2) By a pressure operated valve connected to the refrigerant circuit, i.e. to the compressor delivery or to the condenser. This valve is of the modulating type, opening and closing a little as necessary to maintain a constant condenser pressure. It has the disadvantage that it does not shut off completely immediately the machine stops.

How may several chambers at the same or different temperatures be automatically cooled from one machine unit?

For two, or at the most, three chambers, provided that the difference between the highest and lowest temperatures is not too great—say not exceeding 15°F (8°C)—the liquid line from the condenser may be divided into as many branches as there are chambers and each branch fitted with a solenoid operated liquid stop valve, followed by a thermostatic regulator. Each liquid valve is controlled by a thermostat in the corresponding chamber and the wiring is so arranged that as long as any thermostat is "cut in" the compressor runs. An alternative but more expensive method of control is to circulate brine and control its admission to the various coolers by means of solenoid operated valves.

What precautions must be taken in designing a multiple circuit plant controlled by automatic liquid valves?

(1) Evaporator surfaces for the lower temperature chambers must be ample, whilst those for the higher temperature duty must be limited so that there is no danger of their absorbing all the refrigerating duty available when all the liquid valves are open.

(2) It must be borne in mind that the evaporation temperature and the refrigeration output will fluctuate according to the number of valves open, and the running time of the

machine should be calculated, firstly, on the assumption that one valve only is open at one time, and secondly, with all valves open until one duty is completed; then two valves, and finally one. The mean of the two results should be taken as the estimated running time.

What is an oil-failure relay?

It is an oil pressure switch inserted in the compressor lubricating system and wired to shut down the machine in the event of an oil failure. It is evident that at the moment of starting a machine the oil pressure within it will be nil and the oil-failure relay will therefore prevent starting. To overcome this a timing device is incorporated making the relay inoperative for the first few seconds of starting up, i.e. until, if all is well, the oil pressure has had time to build up to its normal value.

18

REFRIGERATED TRANSPORT

What is understood by refrigerated transport?
The following come within this category:

Small boxes containing material which needs keeping cold during delivery.

Larger containers suspended from the shoulders of salesmen, i.e. ice cream vendors etc.

Insulated containers on tricycles.

Insulated vans and lorries.

Insulated traffic containers.

Rail freight wagons.

River barges.

Ships are not generally regarded as coming under this heading because their refrigerated holds are so large and their refrigerating plant so big that their problems are similar to those encountered in land installations.

What are the requirements of refrigerated transport?
That they should maintain the substances being transported within defined temperature limits throughout their journey.

Would a simple insulated container achieve this?
Not unless the journey time is very short and the container precooled. It depends, of course, on the temperature rise that can be tolerated in any particular case but nothing greater than a couple of hours or so as a general rule.

What are the usual methods of refrigerating transport?
The use of ice, frozen carbon dioxide (CO_2), the direct expansion of liquid nitrogen through sprays, small condensing units carried on vehicles and holdover plates.

When can ice be used?
Only when the temperature required is above freezing and the journey time does not exceed the time taken for the ice to melt. There is a use for ice particularly where the smaller amounts are concerned and it is important that the temperature is not allowed to fall below freezing. The disadvantage is that provision must be made to carry away the water resulting from the melting ice.

When can frozen carbon dioxide be used?
Frozen carbon dioxide (CO_2) is supplied in blocks about 18 in (450 mm) long × 6 in (150 mm) square, and has the great advantage that it passes directly from the solid to the gaseous state without passing through a liquid state. Not only is frozen CO_2 very much colder than ice but its use avoids the necessity for drainage arrangements. There is a disadvantage, however, in that CO_2 will not support life. This means that it should not be used in the holds of barges or in containers having an entrance only from the top, because of the difficulty in properly venting the CO_2 gas before unloading can start. Frozen CO_2 is used considerably in ice cream vending by tricycle or chest box.

What are holdover plates?
These are somewhat similar to the electric storage heater in that they store "cold". Usually containing a liquid or some other substance, these plates are reduced in temperature before a journey and retain enough "cold" to provide the refrigeration required during that journey—provided of course that the journey time and holdover plate capacity are properly equated. The holdover plates may be charged with "cold" by plant at the depot or they may be fitted with their own small condensing unit which can be plugged into the electric mains at any convenient stop, overnight for instance, on a long journey.

What is the alternative to the holdover plates or solid CO_2?
To carry on the vehicle a separate condensing unit. This has the advantage that there is no limit to the length of journey

or duration of stops, provided only that there is sufficient fuel for the condensing unit.

The units are usually driven by a small diesel or petrol engine, sometimes by an engine operating from bottled gas, and sometimes direct from the vehicle engine by means of a hydraulic power generator and hydraulic motor. These are completely automatic and in some cases are fitted with an alternative electric motor drive which can be operated from any public electricity supply; this avoids the need to use the diesel or other engine whilst stopped for the night or weekend.

What insulation is used on refrigerated transport?
The insulation and the entire construction of the vehicle body or container must be rigid, robust and light, particularly the latter which affects the pay-load. These days it is usual for a polyurethane foamed insulant to be used sandwiched between alloy or steel plates. Some designs use a one-piece pressure moulding of glass fibre reinforced plastic with an insulating polyurethane core.

How is the separate condensing unit usually mounted?
It is not unusual for the complete cooling unit to be made as a single item. This is mounted on the insulated container with the front part protruding and the rear part actually within the insulated space. In the rear portion is the evaporator and circulating fan whilst in the front projecting part is the condensing unit with its prime mover.

How is direct nitrogen spray used?
Simply by discharging nitrogen in liquid form through suitable nozzles situated immediately over the part to be cooled. For a fuller explanation refer to page 17. This is a new but rapidly developing method of short term refrigeration, since clearly, the journey time must lie within the capacity of the nitrogen carried. Liquid nitrogen is supplied in steel cylinders and since these themselves are weighty there are clear limits to the system. An advantage is the simplicity and low initial cost.

19

COLD STORE DOORS AND MISCELLANEOUS

What precautions should be taken when first starting up plant?
The details will naturally vary with the size and make of the plant, but the following are of fairly general application:
(1) Make sure that the compressor discharge stop valve is open.
(2) Make sure that the oil level is correct and that the gauge valves are open.
(3) Turn on the condensing water and the jacket water, or start the condensing water pump. Where there is one, start the water cooling fan or condenser fan.
(4) Open the by-pass thus putting the suction and delivery of the compressor into communication.
(5) Start the circulation of brine, water and air.
(6) Start the machine.
(7) When running at full speed, close the by-pass and change over to normal suction from the evaporator.
(8) Open the liquid stop valve.

What other precautions may be necessary on starting if the evaporator is warm and the motor sized for normal working?
It may be necessary to limit the horsepower taken by operating the duty control valve, or simply throttling the suction. It is most important on automatic plants that the compressor motor should be of sufficient horsepower to start the machine against the highest evaporator temperature that could possibly occur.

What points may require attention during running?
The regulator if hand-operated, lubrication and the crank-

shaft gland. Fully automatic plants of small and moderate size should be sufficiently reliable without these attentions except at fairly long intervals.

What kind of lubricating oil should be used in the compressor?
Never ordinary engine oil. The specification of oil to be used is highly important: it must not be too viscous at low temperatures or too fluid at high temperatures and it must retain its lubricating properties at all temperatures to which it will be subjected. In addition it must not react adversely with the refrigerant in use or attack any material with which it may come into contact—not only metals but jointings and packings. The manufacturer of a plant always specifies the lubricating oil to be used and this recommendation should on no account be departed from.

How is an ammonia plant tested for leaks before charging?
By breaking a joint at the compressor suction, closing the suction stop valve, and, with all other valves open, pumping in air until the pressure throughout the circuit reaches, say, 200 lbf/in^2 ($1{\cdot}38 \times 10^6 N/m^2$). The refrigerant circuit should hold this pressure if it is free from leaks.

What precautions should be taken in doing this?
To watch that the delivery temperature does not rise too high; if it does, the compressor must be stopped and the plant allowed to cool. Condenser water must be running.

When the plant is under air pressure how is a leak detected?
By the application of a lather of soap and water to suspect points; bubbles will betray the presence of a leak.

How is the air removed before charging?
By breaking a joint between the compressor and the delivery stop valve—unless a special plug has been provided for the purpose—and, with the delivery stop valve closed, running the compressor. After the plant has been evacuated as completely as possible, it should be allowed to stand for a few hours in order to test it for inward leaks.

How is the plant charged?
The refrigerant bottle should be inclined with its valve downwards and the charging pipe connected between the bottle and a charging valve which should be provided at the inlet to the evaporator. After a pressure of, say, 50 lbf/in^2 ($3{\cdot}45 \times 10^5 N/m^2$) is reached in the evaporator (in the case of ammonia) the liquid stop valve should be closed and the compressor run to transfer the charge to the condenser. Charging should be continued in this way, allowing the refrigerant to flow back into the evaporator from time to time. The charging should be completed whilst the plant is actually working.

What are the symptoms of undercharge?
Suction superheat is higher than normal. If automatic regulation is in use the evaporator gauge is higher than normal and the condenser gauge lower than normal. The evaporator duty is low. If a sight glass is fitted in the liquid line, bubbles will appear in it.

What are the symptoms of overcharge?
If a high pressure float is fitted the plant will "run wet", i.e. the liquid refrigerant will be drawn into the compressor causing knocking, low delivery temperature and reduced evaporator duty. Little or no superheat will be observed. If the regulator is of the low pressure type the excess of charge will be transferred to the condenser, restricting the surface available for condensing and causing high condensing pressure, high power consumption and reduction in evaporator duty. If a hand regulator is used and is adjusted to avoid "running wet", or if the regulator is of the thermostatic type, the symptoms will be the same as in the case of a low pressure float regulator.

What are the indications of air in the system?
Condenser gauge and delivery temperatures both above normal. If the compressor is stopped and the condenser water is kept running the condenser gauge should fall to a pressure corresponding to the water temperature. If it remains appreciably higher, then air is present.

How is air removed (or purged) from a refrigerating system?
Air in the system passes through with the refrigerant until it reaches the condenser, where it is trapped by the liquid seal. The gas above the liquid in the condenser therefore consists of a mixture of refrigerant vapour and air. By stopping the machine and allowing this gas to escape through a purging valve, most of the air can be swept away. It must not be supposed, however, that the air and refrigerant vapour can be separated; they are completely mixed and the air cannot be removed without the loss of some refrigerant vapour. This loss can be minimised by allowing the vapour to pass into a special purging vessel in which it is partially condensed by a coil at evaporator temperature, and the density of the remainder thereby reduced.

What is probably the most frequent cause of trouble in refrigerating plants?
The presence of dirt, scale, etc., and in the case of methyl chloride and the fluorocarbons, of moisture.

What is the effect of moisture in the system?
It may freeze at the regulating valve if the evaporating temperature is below freezing; further, acids are produced which result in corrosion.

If the delivery temperature is higher than normal but the condenser gauge is normal or below, what is the probable cause?
The compressor valves and especially the delivery valves are probably leaking.

What precautions should be taken if it is necessary to pump the refrigerant into the condenser, the evaporator being either the shell and tube type or in the form of a water cooler?
The water must be drained from the evaporator to avoid the danger of burst tubes.

What happens if a thermostatic regulator is installed outside the storage chamber while its bulb is attached to the suction pipe inside, and the regulator becomes colder than the bulb?

As with a thermostat the charge may transfer to the regulator body and the regulator will then close.

How may the furring and scaling of condenser tubes be avoided?
The deposition of scale on condenser tubes is met in recirculation systems when the natural salts in the water being circulated get progressively greater until the concentration becomes such that salts are thrown out.

The usual method of avoiding this trouble is to "bleed" the water circuit so that part of the recirculated water is drained off and replaced with fresh water. The amount of bleed will vary with circumstances and the chemical make up of the water and should be adjusted until the deposition of scale ceases. About 1% of the amount being circulated is about usual.

Very severe scaling can often be appreciably eased by the use of softened make up water, or demineralised make up water.

What is meant by the term "condensing unit"?
The name *condensing unit* is given to an assembly consisting of a compressor and condenser with associated equipment. It is, in fact, a complete refrigerating plant less the evaporator.

What is an Aquastat?
Aquastat is the trade name for a static electric water treatment unit; it has nothing at all to do with automatic control as has the thermostat or pressostat.

What are the requirements of a cold store door?
Effective insulation, lightness, robustness, fire resistance, ease of opening and shutting and the ability to open from either side.

How are these requirements obtained?
There should be only sufficient insulation to prevent sweating and condensation on the outside of the door and the insulant should be as light as possible. A door is subject to

very rough usage and its construction must be very rugged with the insulant usually sandwiched between two timber surfaces. Sometimes a foamed *in situ* insulant can be used bonded to a fireproof backing and/or facing. A lot depends upon the temperature differences and the location of the door; a public cold store door is likely to suffer much more and harder treatment than would a similar door in private ownership. Ease of opening and closing can be obtained by the use of first quality ball bearing hinges and skilled door fitting. The lighter the finished door the better and that is why the insulation is often only that necessary, as stated before, to prevent sweating. There are cases however where, perhaps because of the large size of the door relative to the whole, the resulting heat leakage rate cannot be accepted and in these cases more insulation has to be provided leading to larger and heavier frames, etc.

What types of door are met with?
Doors can be manual of the single or double leaf type; they can be automatic and sliding, either vertically or horizontally.

How is an air-tight seal obtained?
Usually by a simple overlap of the door and frame thus obtaining a flat surface of contact for the seal to mate upon. The seal surface is often fitted with electric heat tracing elements to prevent the joint from becoming ice-bound.

What is an air curtain?
This is, as the name implies, just a curtain of air across a doorway. When fitted to a cold store door it permits free entrance and exit without having to open and close the main cold store door each time.

When a cold store door is open the heavier cold air inside falls out at the bottom of the door—the flow can be felt and it can be measured by an anemometer—and warm air takes its place. The velocity of the air falling out depends upon the pressure difference between the cold air inside and the warmer ambient air outside, and this is different with the different seasons of the year, being greatest in summer. This fact is important because the air curtain is designed to direct a stream of air down the face of the opening and slightly inwards with a pressure which should be equal to that of the cold air inside trying to fall out. If these pressures are equal there will be no loss of air from the cold space, and consequently no warm air coming in to condense and freeze. If the pressure is too much or too little there will be condensation and ice formation inside and/or just outside the door. The discharge adjustment is therefore important and needs to be altered with the seasons of the year, or if there is any significant change in the external ambient temperature.

It follows that the air curtain will not be effective if exposed to strong winds and the openings on which they are used should have wind protection. Where this is not possible another type of air curtain may be used whereby a powerful stream of air is blown sideways across the opening passing between the duct openings on either side. With this type no adjustment is required for seasonal changes but they are very much more expensive.

Air curtains are usually started and stopped automatically by micro-switches in the door frames arranged to start the fan whenever the door is opened. This control circuit is best arranged at a low voltage, particularly where low voltage gasket heating is also installed adjacent to the micro-switches.

What are crash doors?

These are usually met with where temperatures above freezing are being kept. These light swing doors are placed inside the insulated door proper and perform much the same function as an air curtain but are less expensive. They are, in general, not suitable where low temperatures are being maintained, the losses through them being unacceptably high.

These doors must be light yet robust; they are essentially a "bash" door, hit by trucks coming in or out and closing automatically afterwards. They must be free swinging and not cause damage to the trucks hitting them; rubber is a usual and satisfactory material. The provision of vision panels is most essential if accidents are to be prevented.

When are automatic doors usually used?
Generally an automatic door comes into its own where very large and therefore heavy doors are concerned, such as those large enough to allow a lorry or a fork-lift truck to pass right into the store. Automatic operation takes the manual effort out of opening and closing and is a saving of time, since they can be made to be operated by the approach of the truck or vehicle. Automatic operation does not increase the actual speed of opening as compared with manual. In fact it could well be slower, but they do possess considerable advantages in the right circumstances.

How do automatic doors operate?
By electric door engines, pneumatic engines or oil hydraulic cylinders. The operating mechanism is best kept outside the cold space.

How are they controlled?
By a variety of means: a simple manual switch, photo-electric beams, floor pads or a radio transmitter fitted in front of selected vehicles. With the last method only authorised personnel can then gain admission to the store.

How is the duty of a refrigerating plant usually specified?
In Btu/h, or alternatively in tons refrigeration. Where the metric system is in use the kilocalorie or frigorie is used. (Refer also to page 9.)

What are the "standard" conditions for rating refrigerating machines?
The *American Standard Ton* rating conditions are almost in universal use; they are 5°F (–15°C) evaporating temperature and 86°F (30°C) condensing temperature. Where other conditions are related to a refrigerating duty they must always be clearly stated, otherwise the duty is assumed to be at the standard conditions.

How may the evaporating duty of a compressor be measured?
Several methods are available, of which the most usual is still probably to circulate water or brine through the evaporator at a measured rate, observing the inlet and outlet temperatures. A simple calculation gives the heat removed in the evaporator. If water is used, as it might be for evaporation temperatures above freezing, it may be run to waste, whilst in the case of a brine, this must be recirculated through a heater, the input of which must be adjusted so as to balance the heat load exactly.

What precautions must be taken when using the above method?
The flow of water or brine must be adjusted so as to give a fall in temperature sufficiently great to be measured with reasonable accuracy, usually not less than 8°–10°F (4°–5°C); temperature readings must be taken with a reliable thermometer to the nearest tenth, or preferably where °C are concerned, twentieth of a degree.

Care must be taken that the true brine or water temperatures are read and preferably the instruments used should be immersed in the liquid. A sufficient number of readings at, say, 10 or 15 min intervals, must be taken to ensure that perfectly steady conditions are reached. A small fluctuation in brine inlet or outlet temperatures does not matter, but so long as there is a definite upward or downward trend in temperature the readings cannot be considered satisfactory. Due allowance must be made for heat losses to the evaporator and for the heat input to the brine pump, especially when small evaporator duties are being measured. Separate experiments should be carried out to determine these losses. It must be realised that the accuracy of the condenser and evaporator gauges is as important as that of the thermometer. All instruments should be checked before the start of the test and again at intervals if a series of tests is being carried out. The temperature of the liquid just before the expansion valve must be observed and this is particularly important with carbon dioxide and the fluorocarbons. The suction temperature must also be observed carefully; if the suction superheat is a few degrees higher than normal this has little effect, but it must not be lower. It must be realised that at low evaporation temperatures a thermometer in the suction vapour may indicate several degrees of superheat and yet liquid may be returning to the compressor, possibly trickling along the walls of the suction pipe; hence for best evaporator duty quite a high superheat will be indicated. The evaporation temperature must be steady and for this reason a thermostatic regulator should not be used for duty tests.

INDEX